MARKETING FUNDAMENTALS
for Engineers

Stan Haavik

Professional Publications, Inc. • Belmont, CA

How to Locate and Report Errata for This Book

At Professional Publications, we do our best to bring you error-free books. But when errors do occur, we want to make sure you can view corrections and report any potential errors you find, so the errors cause as little confusion as possible.

A current list of known errata and other updates for this book is available on the PPI website at **www.ppi2pass.com/errata**. We update the errata page as often as necessary, so check in regularly. You will also find instructions for submitting suspected errata. We are grateful to every reader who takes the time to help us improve the quality of our books by pointing out an error.

Marketing Fundamentals for Engineers

Current printing of this edition: 1

Printing History

edition number	printing number	update
1	1	New book.

Copyright © 2006 by Professional Publications, Inc. All rights reserved. No part of this publication may be reproduced, stored in a retrieval system, or transmitted, in any form or by any means, electronic, mechanical, photocopying, recording, or otherwise, without the prior written permission of the publisher.

Printed in the United States of America

Professional Publications, Inc.
1250 Fifth Avenue, Belmont, CA 94002
(650) 593-9119
www.ppi2pass.com

Library of Congress Cataloging-in-Publication Data
Haavik, Stan.
  Marketing fundamentals for engineers / Stanley Haavik.
    p. cm.
  Includes bibliographical references and index.
  ISBN-13: 978-1-59126-059-2
  ISBN-10: 1-59126-059-0
    1. Manufactures--Marketing. 2. High technology--Marketing. 3. Engineering--Practice. 4. Engineers--Vocational guidance. I. Title.
  HD9720.5.H33 2006
  658.80024'62--dc22

2006043189

Table of Contents

Preface and Acknowledgments v
1.   Marketing and Engineering 1
2.   What Does Marketing Do? 9
3.   What Is a Market? 21
4.   Figuring Customers Out 33
5.   The Dynamics of Market and Company Growth and Decline 49
6.   Engineering Insanely Great Products 63
7.   Managing Risk in Projects 81
8.   MARCOM: The Art of Communicating with Customers 95
9.   Moving Products Through the Channel 109
10.   Self Marketing 117

References 125
Index 127

Preface and Acknowledgments

This book is for engineers who want to know how the technical products they create are marketed and sold. It focuses on the development of marketable products and on the characteristics of the customers for those products. It discusses how markets are created and how they can decline, as well as many other factors that are important for an understanding of marketing. Readers who want more detail can explore the many references cited that fully represent the core ideas and, in many cases, are the original sources of these seminal ideas.

I have spent most of my career developing and marketing technology products. I was always committed to an engineering career because I liked to build things. But as my career progressed and my interests broadened, I found that I was equally interested in the people who bought technology products and why they bought them. This became more than a casual interest when I moved into marketing jobs, and my career now vitally depended on what I knew about reaching customers.

What I found was that in the technology field very little was known about marketing. Sure there were blockbuster products that seemed to sell themselves: products like the Apple II and Google, leaders in PCs and Search, though 30 years apart. Clearly these products struck the right chord at the right time with huge populations. But what about the products I was working on in large corporations? It was hard to find the market for some of these. It wasn't easy to understand the buyers and their needs, or to cope with the continual industry change that made customers as uncertain about what to purchase as we were about what to supply.

Even at Harvard Business School, which I attended in the early 1970s, the emphasis was on extrapolating sales curves of products from the past, not on trying to understand developing products for new markets using new technology. In Harvard's defense, I gained considerable experience there in thinking about business situations using the case method, and this was helpful. But no one had put together a body of knowledge about incorporating ever-advancing technology into new products, displacing older products, or establishing new markets.

When I found Everett Roger's book, *Diffusion of Innovations*, in the late 1980s I thought I had found the Holy Grail. Here was a man who understood the issues and developed marketing techniques, not in the midst of Silicon Valley, but for use with farmers in the Midwest in the 1930s. Better yet, although many case experiences were part of his book, he was a scientist and recorded and analyzed the results of his experiments with a scientist's eye for rigor. The result was useful guidance regarding methods for thinking about establishing new methods and new products in new communities of interest—what later came to be known as high-tech marketing. Also, in the 1980s and into the 1990s other books were written to address the high-tech issues. These became valuable resources for me. Even Harvard finally got into the act with the valuable works by Clayton Christensen used here in the chapter on market creation and decline.

During all this time I was working in high-tech jobs and I even began a company, Data Communications, where I developed one of the first electronic infor-mation kiosks (which have now become widely used for information retrieval in public places within the United States). When I traveled in Europe in the 1990s I found such kiosks being successfully used in train stations in Italy, whereas in the U.S. they had been only infrequently used to that point. An unanswered marketing question I wrestled with: Why were they used in Italy but not so much in the U.S.? I also worked for many years for Digital Equipment Corporation, certainly one of the most successful high-tech companies during its era. Here I gained a variety of sales and marketing experiences in the midst of the growth of the computer era, but before the advent of

PREFACE AND ACKNOWLEDGMENTS

the internet. However, even mighty DEC made a few mistakes in the rapidly moving technology world, and was brought down by these errors of judgment. This again led to some unanswered questions about high-tech marketing. Later, Christensen's book provided some ways of thinking about how mighty corporations with intelligent men at the helm can be brought down by upstart companies with new technology and approaches. DEC's collapse was a great tragedy, because it had provided great careers for many, many people in an enlightened atmosphere of employee growth and development.

I am indebted to the outstanding and original work reported in the bibliography that has been created by individuals in academic life and those working in industry. Their work has represented an increasingly accurate chart into unknown terrain, the understanding of technology marketing. These books and findings have provided ever more detailed evidence of what to do in our rapidly changing world to successfully market technology products.

I would also like to express my appreciation to Tom Roan of International Data Corporation, who provided many helpful insights in the course of his association at Internet Commerce Corporation, and to many associates at both General Electric and the Digital Equipment Corporation, who were sounding boards for many of the ideas that took their early expression in my experiences at these companies. I also thank my advisor at the University of Rochester (later at Cornell), Dr. Herbert B. Voelcker, Jr., who provided an expanded view of engineering and its relevance in my studies there. He also was an outstanding correspondent during my service in Vietnam providing intelligent engagement while I was in a hazardous occupation. Also, my sincere appreciation to Peter Holm and Heather Kinser at Professional Publications, Inc. who have provided the editing guidance vital to a successful publication and to Miriam Hanes (typesetter) and Amy Schwertman (illustrator).

Technology products continue to be an ever-important part of our developing world. We look to technology both to enhance our lives and to save our world from itself. Solutions for global warming, the apparent depletion in available oil reserves, better means of getting people to work, and automobile safety will all require new technology approaches. However, coming up with the technology is but one part of the solution. The other part is marketing that technology so that it becomes accepted in the broader world. This requires marketing knowledge that is very useful for the engineer and technology professional to know. This book takes the technologist's point of view in providing an introduction to that subject.

Stan Haavik

PROFESSIONAL PUBLICATIONS, INC.

1

Marketing and Engineering

Engineers need an understanding of marketing to design and develop products that succeed. They need a sense of how marketing helps products thrive and how markets operate. Engineers can then anticipate the needs of marketing in the design of the product and the elements that support the product. Engineers who can build marketing's foresight into products will help their company compete successfully. As a friend once remarked, in the competitive world of today you need "to bring a gun to a knife fight." Marketing knowledge can provide the crucial additional firepower needed for successful products.

Marketing is the art and science of predicting and developing demand for goods and services, and of delivering the goods to the right place at the right time. It encompasses product creation, market understanding and definition, communications about the product, and methods of selling and logistics.

Marketing thought and action must mirror the complexity and pace of the world around us because it reflects human culture and the advance of technology.

Marketing was not always so important. In the post-World War II years, American manufacturing companies produced goods that everyone wanted. Pent-up demand fueled high growth. Almost everything made was readily sold to a waiting and enthusiastic audience. Television, radio, and even the telephone were still in their infancy, so instant communications had not made a significant impact on marketing methods. The idea of marketing over the internet or on-line information services was beyond anyone's conception. To put it bluntly, marketing took a backseat to pure production…grinding out the goods was the favored corporate focus. There wasn't really a high level of business concern for marketing or for building products that truly met customer needs because it wasn't needed. People bought whatever was available.

So why is marketing considered so vital today? The global economy creates more competition among firms from all nations. New media have increased awareness of new products and services, generating demand. Increased air travel, tourism, lowered trade barriers, and a relatively peaceful world have promoted trade. The acceptance of greater diversity in cultures and lifestyles provides the impetus for new and alternative products aimed at new niches. For companies to reach these new audiences, with the new media, in new ways, requires fresh ideas and high levels of interest and accomplishment in marketing.

The changes mentioned above have also led to fractures in what were considered longstanding marketing practices. For example, once stable relationships between buyers and sellers have been subject to disruption. In the past difficulties in the trading relationship were often overridden by the strength of personal relationships and the tradition of the companies involved in the relationship.

However, there is not as much patience between long-term customers and suppliers when product problems arise as before.

1 MARKETING AND ENGINEERING

Case Study
Chip Defect

Take the defect in the Intel Pentium chip that occurred in the 1990s. News of the flaw spread worldwide in days, if not hours. IBM—a major, long-term customer of Intel's—pulled back on its support of the chip, which was a major factor in Intel's decision to compensate consumers for the flaw in the chip.

Case Study
Ford vs. Firestone

Ford Motor Company and Firestone had a long-standing relationship (over many years) that dissolved in acrimony over what eventually appeared to be defective tires delivered to Ford for the Explorer SUV. The problem was that the Explorer SUV was involved in a number of accidents where the vehicle actually turned over, causing injuries and in some cases even the death of occupants. The acrimony over who was to blame went on for many months and the dispute was widely reported in the press. Ford claimed that defective tires were the problem, and Firestone

Figure 1.1. Ford vs. Firestone: A Breakdown in a Long-Term Relationship

Printed by permission of MinneHa!Ha! Studios, P.O. Box 6626, Minnehaha Station, MN 55406.

PROFESSIONAL PUBLICATIONS, INC.

claimed that the Ford's design was inherently unstable—leading to a tendency for it to turn over. Regardless of who was to blame, the fusillades of accusations destroyed the longstanding relationship between the two companies.

Regardless of a company's past successes or overall reputation, any slip is quickly amplified and reported in the news media. While these events are very unfortunate and can even lead to the demise of firms, one of the additional responsibilities of marketing can be communicating with the media in helping to mitigate these large problems.

THE TWO BASIC FUNCTIONS OF BUSINESS

Peter Drucker, a professor at Pomona College in California, has long been recognized as the pre-eminent sage of management consultants. He has written many books on management, a number of which focus on the issues of innovation in encouraging growth. In one of his most successful books, *Management*, he wrote

> Because its purpose is to create a customer, the business enterprise has two—and only these two—basic functions: marketing and innovation. Marketing and innovation produce results; all the rest are 'costs.'[1]

Wow! Those are strong words from the man who is arguably the most famous of all management gurus. They point to the fundamental importance of both marketing and innovation in creating value for a company.

The essence of engineering is innovation. Engineering creates the products and processes that drive the growth of a company. According to Drucker, for companies that produce technology-based products, engineering is one of the two business functions that produce results. This is not to say that only engineering can innovate, because certainly there are financial, marketing, and human resources innovations among the functions of a company. But for companies that create products through engineering, engineering and technical innovation is crucial. The other basic function outlined by Drucker, marketing, is a creative process as well. It is the process of creating a customer.

---

[1] Drucker, Peter, *Management*. New York: Harper & Row, 1974, p. 61.

CREATING A CUSTOMER

To create a customer, marketers must understand the capabilities of the product. But beyond that, they must understand the customer's mind. If the raw material of the engineer is technology, the raw material of the marketer is customer thoughts and perceptions. Marketers work to prepare the customer's attitude to be receptive to the product. They do this by assessing whether the product meets key and essential customer needs. Customer concerns that lie hidden beneath the surface must be brought to light and fully dealt with. The marketer must include considerations for the working environment into which the product will be inserted, the customer's views toward adopting new technology, the opinions of all the decision makers who influence purchase of the product, and many other factors.

The process of creating a customer should start at the earliest stages of a product development effort. At this point marketing and engineering should review marketing research studies of what the customer wants, and hold think tank sessions to hash out all the views important to building the desired product. Marketing and engineering want to establish a creative and cooperative tension between their activities. By focusing on customer's needs, marketing imposes a discipline on engineering that helps force the creation of a product valued by the customer. By explaining what is possible to make, given constraints of time, materials, and budget, engineering helps marketing create customer expectations in line with what can be delivered. An investment of time and energy in mutually planning and understanding product requirements will reap benefits many times over as the project proceeds.

After completing the engineering development activity, the company may plan several alpha and beta test releases. This series marks the beginning of the product launch process, in which the focus of activity shifts from engineering to marketing. Customer creation activity now shifts from warm-up to high heat when a real product becomes available. Ideally, marketing would have played a key role in helping to define the product at the beginning of product development. Once the product is agreed, engineering develops it, and then once again marketing moves into the lead to bring it to market.

The company usually discovers during the launch stage that its meticulous prior planning was not perfect. As customers are exposed to the product, the sales department reports that some customers want features that are not in the product. Frantic discussions and mad scrambles may ensue to adapt the product to incorporate some of the newly discovered desirable features. Engineering, which had been looking forward to a well-deserved period of rest, finds itself doing more than just product support during this early stage of the product's life. It is pressed into service to add features and functionality needed for that big sale.

MARKETING—THE SOFTER SCIENCE

Engineers may tend think of marketers as practitioners of a (very) soft science. An engineer might reason that the truly difficult, and therefore worthwhile, activities lie in building things that can be seen and for which the underlying physics, chemistry, and math are understood. Contrast that with a marketer's interests and courses of study in psychology, economics, or business. These courses have none of the precision, discipline, and rigor found in engineering.

However, many engineers who have spent some time in the marketing arena will usually change their assessment about the relative ease of marketing a product. The essential difficulty is that we cannot possibly know all there is to know about how people think and make decisions. The skill of marketing, therefore, lies in assembling and evaluating all the available information to get as close as possible to the true course of action. This course should meet the needs of a satisfactory number of customers and allow room for fluctuations and changes. Marketers then need to direct midcourse corrections as events develop. Marketing totally lacks what many engineers most want: precision![2]

---

[2]It must be acknowledged that, in addition to the precise calculation, there is an art to engineering. In many cases the intuition and experience of the engineer contributes to products which "just plain work," even though educated guesses and so-called "Kentucky windage" enter into the product's design and development. There is an interesting discussion of the motivations and characteristics of engineers in the book *Profession without Community: Engineers in American Society* by Robert Perucci and Joel Gerstl (New York: Random House), 1969. See especially Ch. 2, which discusses the social background and career decisions of the engineering student.

1 MARKETING AND ENGINEERING

Figure 1.2. Marketing plans a course of action and adjusts as needed.

Two Sides of the Same Coin

As companies grow, they divide tasks among a greater number of employees. A natural outgrowth is the tendency to increase the separation between marketing and engineering. However, to remain effective, the two functions must continue to work closely together.

Case Study
Tying Products to Market Demand

An engineer at Digital Equipment Corporation (DEC) once remarked that when the company was small engineers knew customer problems intimately and worked directly with their customers. The solutions to these problems were found and generalized for use in many products. As the company grew, however, engineers spent less and less time with customers. They became removed from solving the customer's immediate problems and more involved with maintaining existing products and resolving internal administrative issues. Marketing was now supposed to be done by product management, which was comprised mainly of ex-engineers. The product manager was supposed to define the products based on three things: how he saw the market, input received from customers, and his knowledge of technical capabilities available in engineering. A layer of management was now inserted between customers and the engineering department, insulating engineers from the true problems faced by the customer. While engineering as a function was as strong and competent as ever, the increasing failure to tie products to shifts in market demand led, many people felt, to DECs difficulties in the early 1990s after the great successes of the 1970s and 1980s.

While marketing and engineering require different sets of skills, they are united in their focus on creating a product that will solve a customer problem. The practice of these respective disciplines often leads to distinctly different priorities because of distinctly different ways of viewing things. Engineering focuses on detail and precise calculation, while marketing provides a holistic awareness of events that affect the market. However, the two have much in common. Both need to be aware of what is on the horizon, either from a technology or customer-needs perspective. Both need to think of possibilities for new combinations of materials and devices, either from the development perspective (engineering) or from the sales perspective (marketing). Both always need to consider the competition.

Success in today's rapidly changing, technology-driven world requires close cooperation between the engineering and marketing elements of any company. They are two sides of the same coin, separate functions but of equal importance. Recognition of the essential differences in approach and thought processes of the two, and a healthy respect and understanding of each, in light of their complementary natures, will lead to more products that succeed in a competitive world economy.

2

What Does Marketing Do?

Skill Builders in This Chapter
- Company Orientation
- Marketing in Your Company
- Changing for the Better

Of course the answer to the question "What does marketing do?" has livened up many a discussion among engineers. While marketing creates some tangible evidence of actual work, such as data sheets and advertising flyers, much of what marketing does can be rather abstract and non-tangible. It is just these non-tangible things, which will be discussed in much of this book, that create the value that marketing can offer. This chapter offers an overview that incorporates an engineering perspective of what it is that marketers do.

Foremost among its activities, the marketing department manages the company's relationship with the customer—the primary relationship for any company. Marketing may or may not be responsible for actual sales; a separate sales department usually performs this function, but some companies include the selling responsibility within the marketing department. However, marketing often develops the method for carrying out the sale. It also designs the way customers perceive or actually think about the company, its products, and its services. Additionally, it helps translate what it understands about customer needs into requirements for products that engineering can design and manufacturing can produce, and influences the creation of services that support the product and the customer.

Communications engineering is the study of the transmission of signals in the presence of noise. Typically, in communications engineering a signal is created at a source and is transmitted over a channel to a sink. Along the way, in the channel between the source and sink, noise is introduced by natural phenomena such as sunspots and electronic devices, and this corrupts the signal so that special techniques must be used to extract the correct signal from the noise. To borrow some terminology from this field of communications engineering, one of the tasks of marketing is to create a channel between the company and its customers. Marketing operates in the channel to create acceptance of products developed by engineering and product development side of the organization. As can be seen in Fig. 2.1, marketing attempts to develop a matched relationship between customer and product. This is directly analogous to the "matched filter" of communications that enables an effective flow of understood data in the presence of noise. Marketing creates a channel between engineering and customer that creates as closely as possible a matching of customer needs with the product produced, and works to filter out the noise in the channel that may be inhibiting the customer's perception of the product.

2 WHAT DOES MARKETING DO?

Some examples of noise that could get in the way of the customer's perceptions are the following.

- information about alternative or competitive products that the customer could buy
- other demands on the customer's time
- lack of knowledge about the product
- competitive actions such as new product introductions, lowered prices, press releases
- internal advice and information from the customer's co-workers

Figure 2.1. Comparing the Marketing and Engineering Models

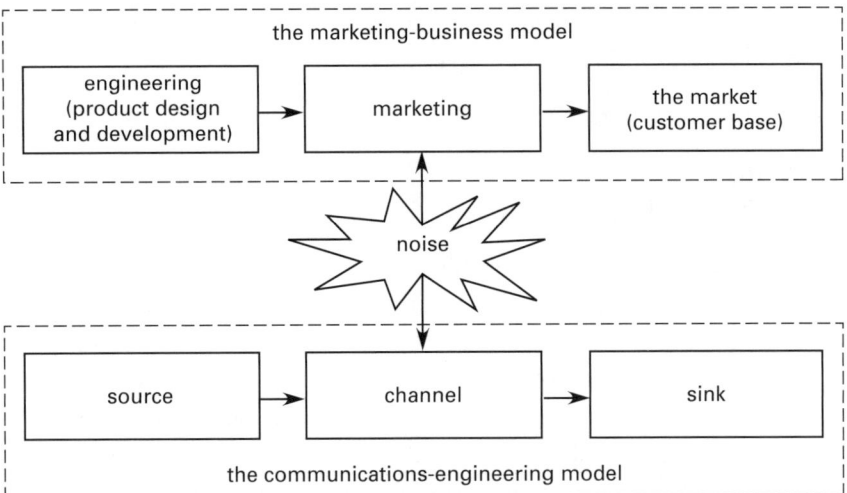

It is not enough for the marketer to know marketing. In order to develop the appropriate and winning messages (communications) that will be received and interpreted by the prospective customers, the marketer must also know the product, and know the engineering that went into the product. The marketer must know the product's technical and user features, and be able to interpret them as benefits to the customer. The language of engineering is features, but the language of marketing is benefits. These benefits must resonate with the customer's needs so that the product will stand out and be remembered in the presence of noise.

In effect, a marketer must straddle two worlds—the product camp and the customer camp. Most marketers never attain, nor do they need to have, the level of product knowledge possessed by engineers, but they must know enough to communicate product features and benefits to each and every one of the customer groups of interest.

Early in their careers, most engineers are firmly grounded in the technology camp. With the rapid pace of technological advances, it is difficult for most to keep up with their own fields, let alone be overly concerned with marketing issues. However, as their careers advance, many engineers find the challenge of dealing with customers occupies more of their time. They become more embroiled in the issue that fundamentally drives any company: product acceptance by customers. If they continue in this direction, they soon find their feet planted in both the technology and the customer camps. Most companies that produce engineered products have engineers who work with customers on a close and continuing basis.

Case Study
The Benefits of Working with Customers

Sometimes engineers must take direct action to clarify, confirm, and deliver what is needed in the marketplace. DEC experienced a string of disappointing product entries to the workstation market in the late 1980s. The workstations of the time were high-performance computing platforms used by engineers for solving technical problems. Sales were bad because, frankly, the products coming out of the division did not meet customer needs. After several engineering managers failed to produce noticeable improvement, a new engineering manager took over the product line. This individual recognized that he needed direct input from the people most familiar with customer needs: the sales department and the customers themselves. The marketing department was failing to supply the needed information, so he usurped marketing's customary role as the interface between the customer and engineering. He met directly with the sales force, which was feeling the brunt of failure, and clarified their view of product shortcomings. The sales force was more than ready to provide a litany of features and capabilities that were missing from DEC's products yet apparently provided by the competitive products. He also met with the customers whenever possible. At each engineering meeting he would evaluate how, why, and in what way the products met (or did not meet) customer needs. His phone was open to calls from the sales force and customers. The new input and attitude recharged a product division that had been dispirited and failing. Leadership like this was sorely needed to build products that met customer needs. There was some improvement in product, but unfortunately the guy got an offer from another company and left DEC before all was turned around.

PROCESS ENGINEERING

Not all engineers develop products. Many engineers develop processes by which products are manufactured. Figure 2.2 shows how manufacturing is often the bridge between engineering and marketing. In fact, in many industries manufacturing or process innovation is as important as the actual product development itself. Process innovation seeks to scale production to economic volumes while maintaining quality and delivery timetables. It is an essential part of a company's competitive capabilities.

Case Study
The Assembly Line

One of the great process innovators was Henry Ford. His introduction of the automobile assembly line produced affordable, quality vehicles for a new car market: the average family. He even helped them afford it by paying wages high enough to enable his employees to purchase it. Talk about priming the pump! By meeting the needs of this poised market for an affordable car, he changed the course of history. His process—the paced assembly line—adapted well to other industries.

Process innovation has also been an important factor in the production of glass, steel, semiconductor, and many other products. Complex yet efficient manufacturing innovations have brought many low-priced, high-quality

Figure 2.2. Idealized Product Flow Through a Typical Company

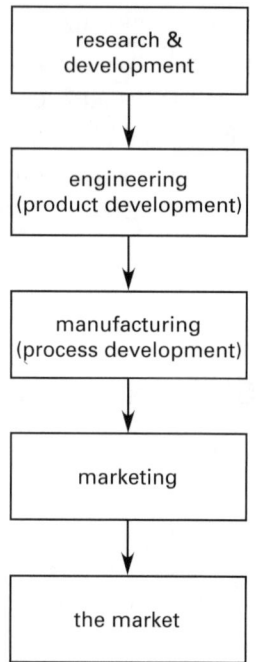

goods to consumers worldwide. Meeting the needs for quality, on-time delivery and product availability in volume are key marketing issues and are a direct result of problems resolved by process engineers.[1]

Figure 2.2 leads one to believe that product engineering and production is, more or less, a process of sequential flow from engineering, through manufacturing, to marketing, and on to the consumer. It is doubtful that this flow method ever worked well, although it was definitely a common model in times past.

In today's faster-paced world, marketing and engineering need a direct, friendly, and collaborative relationship that unites customer needs and product creation. A team approach to product development is now the common model and has led to stunning successes. An example of a team approach is called *concurrent engineering*. This organizational model encourages simultaneous collaboration in the development of products.

---

[1]For an illuminating discussion of the distinctions between product and process innovation, see the book by James M. Utterback, *Mastering the Dynamics of Innovation*, Cambridge: Harvard Business School Press, 1994.

Concurrent engineering seeks to involve the downstream requirements for process engineering and marketing in the engineering design decisions. The reason is very simple: most of the costs of a product are the direct result of early engineering decisions as to how the product is designed. Similarly, sales of the product may be greatly affected by design decisions. If manufacturing only later has the opportunity to review the design, the costs may be locked in and unchangeable. Similarly, marketing should incorporate its views of the product's design before the design is finalized.

The case study below illustrates an engineering department closely involved with the design of processes that enabled the efficient manufacture of the Gillette Sensor razor.

Case Study
Process Design for the Gillette Sensor Razor

> Compared with other razor designs at the time, the Sensor was a highly complex product, incorporating 23 parts (compared with five pieces for most disposable razors). Its complexity created stiff manufacturing challenges. For example, because the blades needed to float on springs independently of one another, they had to be much more rigid (to hold their shape) than those in existing systems and therefore had to be mounted on a thick steel bar. Since gluing them to the bar would have been too unreliable, imprecise, and expensive for high-volume manufacturing, Gillette engineers, working with a vendor, developed a laser spot welder capable of performing 93 highly intricate welds per second—faster than any existing laser. Gillette also needed to develop intricate molds so that it could make the head of the razor (which hold the blades) within tolerances of plus or minus 0.0002 inches.[2]

TAKING THE PULSE OF THE MARKET

An astute engineering organization is always taking the pulse of technology in seeking the most current devices, tools, and methods to incorporate into a

---

[2]Pisano, Gary P., and Wheelwright, Steven C., "High-Tech R&D," *Harvard Business Review*, September-October, 1995.

product. They get this information from a variety of sources: the trade press, attendance at conferences, discussion with colleagues, competitive products, and so on. Engineers can also draw upon suppliers' knowledge of new devices and tools. After all, the suppliers must know a thing or two about what customers want, or why would they be producing these devices? An engineer imaginatively incorporates the newest and best while at the same time understanding and factoring in the market's needs for features and performance—taking the market's pulse. If you've ever tried duck hunting or skeet shooting, you know you have to lead that target as it flies away from you. Similarly, you have to understand the pace of technology and where it is going to create a successful hit. Wayne Gretzky, the great hockey player, once said that he always skated to where he thought the puck would be. While hockey is an extremely dynamic sport with players always in motion, the intelligent estimation of future direction and location of the puck may pay off in a competitive edge.

Case Study
Reading the Market

Consider the iPod, Apple's smash product hit. Before the iPod, there was a market of consumers downloading and sharing music on the web, yet there was no really effective way to transport that music to the gym, the car, the beach, or the bicycle, satisfying the needs of a mobile generation. Further, some methods of file sharing were found to be illegal. Clearly there was a market need being felt and developing. And the MP3 player and other standards came out to meet it. Yet only Apple was able to create a truly outstanding combination of components, engineering design, and design for human use and pleasure, that led to its runaway success. Customer needs must be led in order to have a product arrive in the market when the market is ready for the features. The Apple engineers and marketing staff guessed right about most of the factors essential to the iPod's success.

Compare this to a product development effort in which the engineering staff decided to use a new graphics generator chip that was in beta test by a major manufacturer. The chip offered many desirable new features, but was not bug free. Selecting this chip led to delays while the bugs were solved. This created huge uncertainty for marketing as to when the new product would be available, and it was impossible to predict product performance due to the bugs. Ultimately, the six-month delay of the product, a graphics display generator, created enough time for several competitive offerings to become more attractive. The delay also lowered the ultimate payback from the product's sales. Customer enthusiasm from beta sites seemed to wane as all waited for the engineering problems to be solved.

Clearly, in the case of Apple, there was great awareness of market trends for a new product capability. While the inside story of the iPod's development has yet to be written, one feels that the marketing department at Apple, together with the visionary capabilities of its CEO, Steve Jobs, must have been dynamically engaged in a productive discussion over many of the issues involved in the iPod's development. In contrast with the Apple story, my sense of the chip selection is that the engineering department was sold on advanced features and not on the risk attendant to using a chip that may not be available in time for the market. Consequently engineering and marketing struggled with a commitment that had been made and was too late to reverse. On the other hand, a decision to use a more conservative, existing chip may not have produced the leadership product that was desired.

Just as an engineer is sampling available technology, a marketer is continually taking the pulse of the market. In modern technical markets, the pulse is often not only unsteady but can be subject to wild fluctuations due to new companies that enter promising markets, new product announcements, questions about the appropriate technology to use to satisfy emerging market needs, and the like. In this changing environment, a marketer needs to identify: What is the current market's opinion to actual performance or functionality considerations? Is the market growing or flattening? What is the true market for this product or service? These are basic questions, but many companies do not truly understand or carefully define their market.

Case Study
Understanding Market Changes

> The former president of DEC, Ken Olsen, was a key industry figure whose background was in engineering. His personality inspired a culture throughout the organization based on excellence in product development. For many years the company was a leader in the computer industry, with success after success cementing the company's stellar reputation for technical excellence. Unfortunately, this smoothed road eventually turned rocky as competition stepped in and rapidly took DEC's market share away. This was directly caused by the competition's more perceptive market awareness, as well as excellent product development. As a result of not understanding market changes at a deep level and acting on them, DEC went through years of painful restructuring. Eventually it became so weakened that an upstart PC Company—Compaq Computer—bought it.

An example of a company that has stayed close to its markets and its customers for over 50 years is Harman International—this even though its customers are in consumer electronics, one of the most dynamic, changing, and competitive of all markets. In recent years Harman has achieved a leadership position in supplying advanced electronics and audio systems for luxury cars. Sidney Harman, the chairman of Harman International, talks about his marketing philosophy in Sidebar 2.1.

Producing quality products by itself is not enough. While product quality and engineering talent are a necessary prelude to success, more is required in the

---

Sidebar 2.1: Staying Close to a Market

"We want to be close to the markets we sell to, operating with local people who understand the culture and practices of the local customers," said Harman, the Canadian-born engineer who co-founded the company Harman International in 1953. "Our industry (consumer electronics) has been burdened with a technological determinism—if the engineers can design it, the company builds it. We reverse that...The company's role," Harman said, "is to act as 'anthropologists in the marketplace,' to figure out what people will want."[3]

---

[3]"Washington Business," *The Washington Post*, September 4, 1995, p. 10.

modern world. Companies still succeed on the basis of product excellence, especially in the early stages of growth, but marketing is needed to help the company stay focused on customer needs.

There is another side to the coin. Companies that are all marketing, that lessen the emphasis on product excellence, run the risk, as Regis McKenna says, of being "all hat and no cattle."[4] Such companies may lose their markets through failure to deliver on the product promises that marketing has made.

Skill Builder: Company Orientation

Think about your company. Does it have an engineering/product orientation, a marketing orientation, or is it pretty well balanced?

What is the background of the leadership—engineering, marketing, finance or other?

How does the background of the leadership affect your company's orientation?

Is your company focused enough on the customer?

Skill Builder: Marketing in Your Company

Is the marketing department in your company sympathetic to engineering processes and development issues?

Has a product development effort ever failed because marketing set timetables that were too tight or set specifications that frequently changed?

Skill Builder: Changing for the Better

If you perceive inefficiencies or other needs for improvement in the way marketing interacts with engineering, how would you change the current methods?

---

[4]McKenna, Regis. *The Regis Touch: Million-Dollar Advice from America's Top Marketing Consultant.* New York: Addison-Wesley, 1985.

3

What Is a Market?

Skill Builders in This Chapter
- Markets
- A Job to Do
- Discovery

This chapter is all about markets: what they are, who is in them, and how to ferret them out. For engineers working in technology-driven markets, these aspects of marketing are made difficult by product newness—there is little evidence or history on which to base estimates of future demand for products. One just has to cope with this aspect of the new market and find ways to handle the uncertainty (which is discussed in Ch. 4). Because markets are composed of people, it's easy to envision marketers as anthropologists. They focus on understanding the culture, attitudes, and tools used by groups of people. They seek to identify groups that are potential purchasers of their product or service. They can also be viewed as psychologists because their focus is on the behavior of individuals and how they respond to the messages and information that is aimed at them.

This chapter looks first at how a market is characterized, and then at methods of discovering subcomponents, or segments, of markets.

DESCRIBING A MARKET

Simply put, a market is a collection of people who consult with each other about their buying decisions. They are connected in some way with common interests that unite them around a certain product category. Municipal wastewater treatment plant managers, for example, are interested in the equipment used for processing water. People who wash clothes have a common interest in the effectiveness of their detergent. Given the opportunity, people in these groups will discuss or consult with each other about their choices to solve problems or perform common tasks.

In these groups of buyers or potential buyers that comprise markets, the amount of reference to each other or to other resources about the product purchase will vary. Usually, the need to consult with colleagues or outside sources is directly proportional to the perceived risk a buyer feels with the purchase. Imagine the manager buying software to run a critical business function from a company that is totally unknown to him. He would want to check with other buyers using the software for the same applications, to be sure its operation and short- and long-term support would be reliable.

3 WHAT IS A MARKET? 23

A technical manager in Detroit, for example, interested in buying CAD/CAM (computer-aided design/computer-aided manufacturing) equipment might refer to a CAD/CAM user in Baltimore who was using the equipment for similar applications. This group of CAD/CAM users applying equipment to solve similar problems constitutes a market. Conversely, an engineer in Atlanta using the same equipment for solving different problems would not be in that market.

New products or services are critically dependent on references for additional sales. When a seller has established a number of references in a market he usually encourages communication among the users, hoping to promote more applications for the product by word of mouth. User groups are created out of this philosophy.

What about a product that has a high degree of engineering content but is significantly less risky to purchase? This might be a purchase decision for a power tool, like an electric drill, or maybe a CD system. You might buy such an item on the basis of the recommendation of the salesperson, or on the basis of ads in the newspaper. While such a purchase is less risky, marketers will still define the market as a segment consisting of persons who have similar characteristics. In the case of a CD system, the primary market segment might be all young adults between the ages of 18 and 30. These individuals can be counted on to discuss purchases with their peers and can be greatly influenced through ads in magazines and other media that appeal to their age or other demographic characteristics.

Markets are always described by the people in them. The CAD/CAM market, for example, is composed of users of CAD/CAM equipment or of people who perform functions requiring the use of CAD/CAM equipment. It seems obvious, but we do need to remember that CAD/CAM equipment is not a market; it is a set of products sold into a market.

DIVIDE...

Do you think the entire world will beat a path to your doorstep upon your discovery of the better mousetrap? Neither do most marketers. Instead, what they do is divide things up into more understandable and manageable parts, a process called segmentation. Each segment will owe its existence to the use of certain variables that characterize the persons in the market segment.

The old mnemonic of *Who, What, When, Where,* and *Why* forms an ideal, if timeworn, way to begin market segmentation.

Who

Who are these people that may have an interest in a product? Their age group, educational level, occupation, marital status, race, religion, income, ethnic identity, family, hobbies, and other information may be crucial to know.

What

Do they live in houses, apartments, or condos? Do they own or rent? What kind of cars do they drive, what do they read, what non-essential items do they own?

When

When will they want to buy the product? Do they have a strong need now or later? Do they have a strong desire now or later? Do they have the funds to purchase it now or later?

Where

Where are they? Do they live in cities or suburbs? Do they work in factories or office buildings or at home?

Why

Why would they be interested in the product? What need do they have that it will help satisfy? Is it a necessity or a luxury?

The answers to the above questions can help form a base of knowledge about prospective buyers. In consumer mass-marketing, understanding the demographics of the target segment forms the core of the marketing strategy and tactics. And this information is also useful for engineering products as well,

3 WHAT IS A MARKET?

especially when those products—for example, PCs, consumer software, or tools—are directed at consumers.

In Table 3.1, a hypothetical segmentation of the market for power saws is illustrated. Here is a market undergoing considerable transformation due to the introduction into the United States of a number of brands from Asian manufacturers, as well as increasingly powerful portable (battery-operated) saws. In this hypothetical segmentation, the saws are classified into six different categories depending on portability, price, and type of end use. The demographic classification is on the basis of age, but words characterizing the segment are applied to the age range, suggesting the typical occupant of that age group. Finally, each segment receives a rating based upon the perceptions of the marketer as to the desirability of that segment for the producer of the saws. Missing from this classification is the effect of competition on the desirability of a segment. For example, some segments may be owned by a manufacturer in the sense that they have a very large market share there. In such a case, a new vendor may well choose what would be considered a less-desired segment in order to avoid the strong competition and establish a foothold first.

Table 3.1. Hypothetical Market Segmentation for Power Saws

segment title	age range	U.S. pop. in this range (000's)	portable saw <$50	portable saw >$50	starter model	casual use	semi-pro	craftsman quality
young adult first time	20–29	19,413	A	B	C	D	E	E
homeowner	30–39	19,073	B	A	A	B	C	D
family man	40–54	31,765	C	A	B	A	B	C
independent	55+	29,506	D	A	D	A	A	B
all segments	20+	99,757	C	A+	C	A–	C	D

rating system
A = highly desirable segment for this saw classification
B = attractive segment
C = average attractive segment
D = unattractive segment
E = undesirable segment

...AND CONQUER

Demographic segmentation is a very useful and simple way to obtain a degree of understanding for a new market. In consumer mass marketing, it is generally the main method used in devising marketing campaigns.

What this method lacks is a critical evaluation of the market's *need* for the product, something vitally important in marketing engineered products. While need is addressed in the previously posed *Why* question, there is more to it than that. There are approaches to segmentation and targeting segments based on need. Table 3.2 illustrates an approach based on considering consumer needs.

It is not enough to understand markets in general terms. The above examples of segmentation are relatively lightweight and easy to understand, but that doesn't mean that segmentation is an easy task. Each segment's motivations and desire to buy must be examined carefully. Even if the segments are in the

Table 3.2. Needs-Based Segmentation and Attractiveness

segmentation step	description
Discover needs.	Identify groups of potential customers based on their need to solve a problem, do a task, or obtain a certain benefit. *For example, in the case of the power saws above, a new need could be the need to cut complex angles for interior carpentry work.*
Refine the groups into segments based on similar needs and characteristics of persons in the proposed segments.	Research demographic characteristics, characteristics of usage of existing solutions if any, and other factors. *Persons using a saw capable of setting up and cutting complex angles might be professional carpenters. How many professional carpenters are there?*
Determine relative attractiveness of each segment (i.e., how well does the product solve the customer's problem; how entrenched is the competition; what kind of demand is evident based on available discretionary funds to buy the product).	Evaluate the ease of solving identified problems in each segment with the product and the potential payoff in terms of revenue and profit in each segment. *What alternative saws are available, how well do they do they job, what is the relative cost of the new vs. the existing solution?*

same industry, they may have very different buying characteristics. For example, in the pharmaceutical industry two major segments are the manufacturers of patented drugs and those that manufacture generic drugs (i.e., not protected by a patent). Because the first segment reaps huge profits from their patent-protected products, it could be thought that they would be willing to spend more on new means of drug distribution. However, this turns out not to be the case—possibly because their well-funded relationships with companies that handle customary means of distribution work well enough for them. On the other hand, generic manufacturers are in cutthroat competition to lower distribution costs in order to boost their very thin profit margins. Therefore, even though the generics have less money to spend, they are motivated to spend it on lowering the cost of distribution.

Further, even in some cases where a marketer has done an excellent job of segmentation, markets may not respond immediately to the new product. They may need time to adapt to the use of the new product as it may require workflow changes, changes in relationships, and training among other needs. While the segmentation may illuminate many inhibitors to successful adoption of the product by the segment, it is almost impossible to find or acknowledge all the difficulties that may present themselves. Only by wading into the market, armed with the best possible foreknowledge, can the marketer be able to anticipate as many difficulties as possible and hopefully overcome the remaining ones.

IS THERE A "JOB TO DO"?

Clayton Christensen and Michael Raynor, in their insightful book *The Innovator's Solution*[1], discuss finding markets in terms of a job to do. With any product or product idea, what is the job to do that the product helps the market do? Many products have failed because, even though conceived with much thought and attention to marketing analysis, there was no real job to do.

---

[1]Christensen, Clayton M., and Raynor, Michael E., *The Innovator's Solution*. Cambridge, MA: Harvard Business School Press, 2003.

Case Study
Transistor Transition

Let's take an example of Christensen's thinking that meshes technology, engineering and marketing together. When transistors were first available in the 1950s, many large companies like RCA looked to this new solid-state technology to replace vacuum tubes in the radios and TVs of the day. They invested huge sums of money attempting to adapt the transistors to the needs of TV set production. Unfortunately these investments, while certainly of long-term value in increasing their knowledge of transistors and improving solid-state capabilities, had no near-term payoff. The technology was not on the same level with current vacuum-tube technology, and thus it was too difficult to transform or replace vacuum tubes in that early era.

Meanwhile Sony took the transistor technology of the day and focused it on the hearing aid market, where the low-power, rugged, miniature transistors could be readily used to improve the current devices. Their new hearing aids captured this market, which had been plagued with bulky and inconvenient hearing aid devices of dubious effectiveness. After success here, they then they applied the transistor to something new—the first transistor radio.

As Christensen says, "Compared with the tabletop radios made by RCA, the sound from the Sony pocket radio was tinny and static-laced....Rather than marketing its radio to consumers who owned tabletop devices, Sony instead targeted the rebar of humanity—teenagers—few of whom could afford big vacuum tube radios....The teenagers were thrilled to buy a product that wasn't very good because their alternative was no radio at all."[2]

The lesson here is that technology enabled the discovery and development of a totally new market—teenagers. The job teenagers had to do was listening to their own music independent of their parents. Their own affordable radio facilitated this.

---

[2]Christensen, Clayton M., and Raynor, Michael E., *The Innovator's Solution*. Cambridge, MA: Harvard Business School Press, 2003, p. 104.

Case Study
Creating a New Market with a Job to Do

> Another example of a job to do concerns the development and marketing of the Weed Eater grass-cutting tool. George Ballas, a Houston businessman, was frustrated with trimming the grass around trees and other obstacles on his property. He took a tin can, some nylon fishing line and a motor and brought them together into the Weed Eater, now a prominent tool in most garages in the U.S. After perfecting the invention he created the Weed Eater Company that began its own manufacturing and soon developed a rapidly growing market among homeowners who had a job to do—trim weeds in places where lawnmowers could not go. With its huge success the Weed Eater Company soon attracted competition from many companies who sought to find ways around the product's patent protection. Eventually the company was sold to one of the major companies in the lawn care business.

The idea of finding or understanding a job to do is one of the best means to develop and secure a market where there is currently no market or where the competition is, in effect, non-consumption. In both cases cited, there was no consumption of the product and no similar product. In the case of the radio, a new segment—teenagers—was identified, and marketed to. In the case of the Weed Eater, while clearly the lawn equipment manufacturers knew their market segment, they did not have the imagination to invent this new method of meeting evident needs.

A FOCUS INTO MARKETS

Michael Porter's competitive trilogy—differentiation, low cost, and focus[3]—will be discussed later in this book. According to Porter, at least one of the three must have a strong presence to create a competitive offering. For the moment, however, consider how to apply just the focus part of the triad.

Marketing begins with the discovery of segments that are or will be receptive to your product. When that product competes against existing products, you

---

[3]Porter, Michael E., *Competitive Strategy: Techniques for Analysing Industries and Competitors.* New York: The Free Press, 1980.

must understand what the segments are and turn that knowledge into a marketing strategy to create competitive advantage. Identify characteristics of your product that people in the segment want and that are neither available from nor emphasized by your competition. Focused marketing, capitalizing on segment knowledge, and concentrated effort makes for a winning strategy.

On the other hand, suppose you have developed a product that has no competition, i.e., no existing comparable product. You need to find a segment that will buy this product; or there is a segment that is buying, but you want to find a new, less competitive segment. In marketing language this is sometimes called searching for green space, open fields where there is minimal competition. In these situations, focusing may be very premature. Sometimes there is a waiting period before the market discovers your product. It is as if you put your product on the stage and opened the curtains and waited for the audience to file in—the same audience that heard about, wanted to learn more about, and finally bought tickets to the show. But marketing is not a passive activity. You need to go out and bring those customers in, disparate though they may be, and find out if they are the self-selected segment that will buy your product.

A STRATEGY FOR STRIKES

Consider that in many markets and for many technology products there are multiple applications for the same product. For example the Weed Eater could be applied to large-scale weed cutting or to just trimming the spare grass close to trees. While simplistic, these are two different applications of the same product.

Take a software product used to secure a corporation's intellectual property (i.e., the written material meant for a specific audience). In this case the products could focus on controlling emails or on controlling Word or Excel documents. Different market segments have different priorities for the control of these two different types of documents. It is important, therefore, to choose both the segment and the application correctly.

A strategy for entering a market with a chosen lead segment and application is like bowling for strikes. You want to choose the lead segment and the lead application—that is, the headpin. Once this pin is knocked over, and you've established your product in this market, choose the second pin, which could be the same application in a different market segment or a different application in the same segment. The objective is to minimize the friction between segment and applications, minimize the added resource expenditures, and create focus while promoting the expansion of the product market.

Skill Builder: Markets

What market segments is your company in?

What is the population of customers in the largest three market segments?

What segment is your product sold into?

How many customers are in that segment?

What is your share of that market?

Skill Builder: A Job to Do

What job does your customer need to do?

How well can it be done with your product?

Is there room for improvement in your product?

Is there a bigger job that can be done if the product were modified in some way?

Alternatively, could part of the job be done with a less expensive version of the product?

Skill Builder: Discovery

Do you hear of customers discovering new uses or applications for your product?

Is your company open to encouraging new uses?

What support is available for new uses?

4

Figuring Customers Out

Skill Builders in This Chapter
- Market Research
- Life Cycle Stage
- Relative Advantage

Countless millions of dollars are spent each year researching customers in an attempt to understand what they will buy and why they will buy it. Advertising agencies, market research firms, industry analysts, futurists, visionaries, and a company's own marketing department are part of this large industry that focuses on figuring customers out.

HOW THINGS WORK VS. HOW PEOPLE WORK

Many engineers chose their career because they were fascinated with how things work. Engineering is about building *things* that work. Many have enjoyed tinkering with or building things from an early age.

Marketing is about something quite different. It is about understanding how *people* work—specifically how people make buying decisions—and then using this information to help build and sell products. So it is not only about how people think and operate, but how people interact with things.

Most engineers can (and want) to understand people, just as most marketers can understand how cars run or calculators compute. Some people who built large successful enterprises, people such as George Eastman (Kodak) or Thomas Edison (GE), were obviously great engineers, inventors, and tinkerers. However, they were also well known for astute attention to their prospective markets and the needs of the people who would buy their products.

These 19th century engineers were able to contribute successfully to both the engineering and marketing sides of their businesses. Today, however, engineering projects of any size require large and multifaceted engineering teams. On the marketing side, there is usually one leader who is responsible for the plan and execution assisted by many resources—from outside consultants who may handle public relations and advertising to in-house sales personnel who set up the channels, among many other tasks. Executing the marketing plan can be as complex as the engineering program and often requires significant resources.

MARKET RESEARCH

The first task is to better understand the proposed product's market. These activities go under the general heading of market research.

The market research industry provides a large variety of services. One catalog of market research services listed over 1000 companies offering 94 categories of research services. The type of research offered is divided into two broad groups.

- forecasting buyer behavior
- analyzing why customers bought a particular product

With forecasting, researchers often make educated guesses as to what people will buy based on established trends. Forecasters review buying patterns for similar products, or products in the same market, and predict marketplace performance for the product in question. With analyzing, researchers interview customers to try to understand why they bought, or didn't buy a product, or to discover both the features they wish the product had or those they feel weren't needed.

Case Study
Market Research

> Even when exhaustive research and the best techniques are employed, figuring out what customers will do is inexact and prone to failure. Akio Morita, former president of Sony and inventor of the Walkman, developed the idea for the Walkman because he wanted to have music with him as he walked around. Though many at Sony advised against the product, the Walkman was designed, engineered, and marketed to legendary success.
>
> In other examples, studies were conducted to estimate demand for copying machines and office computers. Though almost no demand for these products existed, development continued despite the negative research. The companies that sponsored the research: Xerox and IBM.

To add to the difficulties that market researchers experience, they can get caught up in the general enthusiasm that sometimes surrounds the introduction of a product. This can lead to faulty estimates of the amount of time it will take for a product to be adopted. Numerous high-tech innovations, such as the AT&T Videophone (which provides a television picture along with a voice conversation), the information kiosk, and neural network software, received great acclaim as they were introduced to their initial markets. However, they soon were lost from public view when they didn't catch on right away.

Some products fall into a chasm after an initial flurry of success and publicity. If they emerge from the chasm, it is because they have gone through a period of evaluation and refining and have been adapted to provide a powerful, compelling advantage to a well-defined market segment. There is more discussion of the chasm phenomenon later in this chapter.

To say the least, the science of determining market demand is not perfect and there have been some significant failures. There is no doubt, however, that it can provide useful insight into markets and into product features that are most desired.

Whether forecasting or analyzing buyer behavior, two schools of thought exist. One method bases predictions on quantitative analytic methods, while the other uses qualitative means to understand buyer behavior.

QUANTITATIVE RESEARCH

The quantitative approach works more effectively when the market for a product or product category has been developed and there is some past numerical basis upon which to forecast sales success. A common technique is trend analysis, which uses past sales figures to estimate future sales. Another technique is to call or email potential buyers and ask them what their buying plans are for a certain product or product category. Either way, forecasting buyer behavior in this numerically oriented way relies on consistency not only in human behavior, but also in the product category and the overall market structure. If an unforeseen event occurs—for example, a competitor introduces an alternative product—all estimates can be wrong.

Because markets can change rapidly, definitions of markets can be different, and research firms can interpret data differently, significant disagreements in market size are frequently noted, as demonstrated in Sidebar 4.1.

> **Sidebar 4.1: Forecasting and Grains of Salt**
>
> Even for well-established markets, as with the market for object-oriented data base systems in 1995, opinions about market size can be quite different. For example, market research firm Ovum, Ltd. claimed that the size of the market in 1994 was $80 million. A different firm, Input, concluded that the size of the market was $190 million, more than twice the size estimated by Ovum.[1] Most forecasts and data about market size must be taken with a grain of salt!

When evaluating the case to build or not build a product for a market, forecasts must be evaluated carefully. Even forecasts that are reasonably similar may be suspect on the basis that the two authors might have discussed their results with each other and thereby lost their independence. It is best to consider forecasts along with as much other available evidence as can be found and draw your own independent conclusions.

QUALITATIVE RESEARCH

Qualitative market research, while lacking in numerical precision, more easily helps project demand for new products. This research is more successful at revealing a buyer's true feelings and attitudes toward proposed or existing products. As with quantitative research, there are a number of techniques. In most of these qualitative techniques, participants are paid a small fee for their time.

The most frequently used qualitative method is focus group research, where a group of individuals with a desired level of knowledge about the product or service is assembled in a conference room. For example, the focus group may be composed of Directors of Information Technology at insurance companies, a proposed target market for the company sponsoring the research. A group facilitator, who has been trained in the focus-group technique, elicits the individuals' feelings and attitudes about the product under investigation. One-way mirrors are frequently used to allow the sponsor to view the discussion without obtruding into the group's thinking. If the group has been well selected and the focus group facilitator is skillful, the sponsor can expect to receive a

---

[1] *OBJECT Magazine*, SIGS Publications, March-April, 1995, p. 10.

wealth of opinions and attitudes concerning the product—information that can be used to assess product features and marketability.

Other qualitative techniques include scenario analysis and the Delphi technique. In scenario analysis, persons familiar with the product are asked to write down actual scenarios for product use. The scenarios are analyzed by researchers to evaluate product features. There is a good example of scenario use and examples in *Crossing the Chasm* by Geoffrey Moore.[2]

In the Delphi technique, a jury of knowledgeable individuals is asked a series of questions about the proposed product and asked to forecast demand. Each person is unaware of who other members are. After each round of estimates is completed, all the individual and summary results are shared among the jury and a new round is conducted. After several rounds the results converge toward a "best guess" for the group and the product being evaluated. An advantage of the Delphi technique is that charismatic or persuasive individuals who might dominate a focus group or other face-to-face meetings have less influence.

With any technique, forecasting new markets can be discouraging. It is not easy to elicit opinions from individuals in the brief time available for such encounters. Qualitative techniques depend on the ability of individuals to project themselves and their usage patterns into situations where they may have no experience. One must rely on the skill of the researcher to properly evaluate the input from individuals. Even when persons confirm a desire for a product or service, further research must be done to discover how much customers would be willing to pay and when the market demand will be large enough to sustain a given level of engineering and support activity for the product.

Case Study
Crystal-Ball Gazing

Despite the drawbacks of qualitative market research, some information is better than none. One focus group conducted research to understand the attitudes of software development professionals toward new tools and methodologies used in software development. The research provided

---

[2]Moore, Geoffrey A., *Crossing the Chasm*. New York: HarperCollins, 2002.

key information about the important software interests of these individuals. As a result it was possible to design marketing literature to catch the market's interest, which then led to successful product sales. In another case, a focus group conducted research to determine the demand among healthcare chief information officers (CIOs) for a new means to transmit important financial transactions over the internet. They determined minimal demand for this new solution so the project was dropped. Years later, the internet became a prime means of transmitting financial information for these companies but, at the time, it was a losing proposition.

CUSTOMER DYNAMICS—THE TECHNOLOGY ADOPTION LIFE CYCLE

One of the most fascinating aspects to marketing technology-based products is the way in which the type of customer that buys the product changes over the life of the product. Many engineers are familiar with the product life cycle. There is a comparable life cycle that models the adoption or selection of products by types of customers. This life cycle focuses on the tendency of a product to be adopted by a customer. This model is called the Technology Adoption Life Cycle, or TALC.

As can be seen in Fig. 4.1, there are three broad market categories in the TALC model.

- early market, comprised of innovators and visionaries
- mainstream market, comprised of early and late adopters
- late market, comprised of laggards

The TALC grew out of work done by the U.S. Department of Agriculture (USDA) in the Midwest in the 1930s. The USDA was keenly interested in how innovations in farming methods would be adopted by farmers and how rapidly these methods would spread among the farming community. Dr. Everett Rogers, a USDA field agent, played a major role in the formulation of the TALC model that has been extremely useful for understanding a wide variety of technology adoption processes. From the adoption of birth control methods in third-world countries to the diffusion of inventions originating in Silicon Valley, the TALC has been applied.

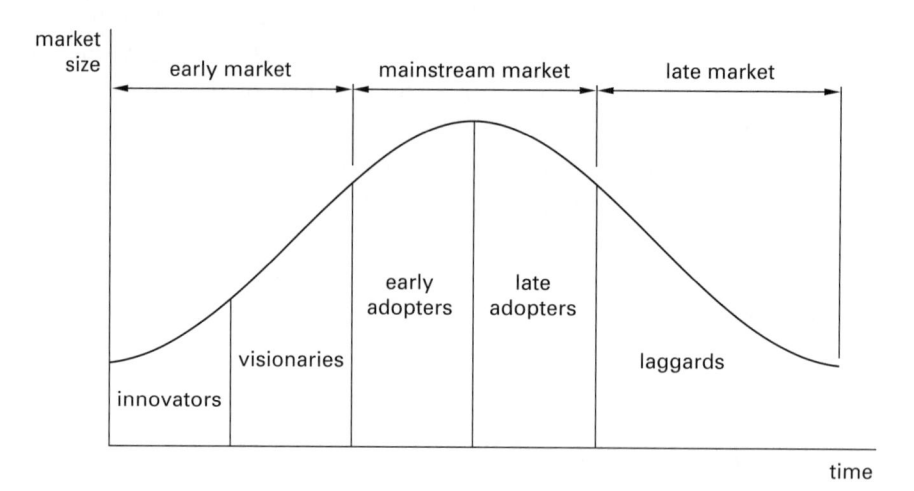

Figure 4.1. The Technology Adoption Life Cycle

The customers occupying each market phase are significantly different in their characteristics and buying attitudes toward technology, so each phase of the life cycle poses significant challenges for marketers and for engineers.

The Innovator

Engineers will probably identify most with the early market Innovator, who functions in many organizations as the gatekeeper to new product acceptance. Innovators are enthusiastic about new technology and are eager to buy or try out the latest piece of new equipment, electronic gear, or software. They read magazines where products are evaluated. They roam the aisles at trade shows looking for the latest and greatest. They try out or evaluate new technology-based products. If convinced of a product's greatness, they will recommend it to the visionary.

Unfortunately, although innovators may have the prescience and technical ability to accurately judge winning products, they don't usually have the organizational clout or funding to buy and adapt the product to the organization's needs. The person with the money is the visionary.

The Visionary

The Visionary is also a technology enthusiast whose agenda is markedly different from that of the innovator. Visionaries are not interested in technology for its own sake, but only for what it can do for their company and their personal careers. Furthermore, the visionary judges whether the new widget will create revolutionary, positive change or only ripples in the corporate pond. Because of the difficulty in bringing new technology into an organization, visionaries feel it's not worth the effort unless major positive change is the result and the visionary will receive credit for the success. The visionary has the ability to win others to the cause of the new technology and can raise the funds necessary to adapt the product to the needs of the organization.

Therefore, to win in the early market, it takes a combination of winning the enthusiasm of the innovator, as well as the pocketbook, commitment, and clout of the visionary.

The Early Adopter

Once the innovators and visionaries are on board, marketers focus on moving the product into the mainstream market. It is in this attempt that many marketing efforts fail because they think early successes translate easily into broad market acceptance. The opportunity to achieve significant market volume and financial success is found in the mainstream market. The Early Adopter is the individual who is the key to success in this market.

Early adopters are usually department heads or higher-level managers who are well connected within their industry. They got to these positions by understanding organizational politics and having a number of well planned, executed, and managed successes under their belt. Before acquiring any new product, early adopters will assess the risk of adoption and likely consider the following.

- Is there a compelling need in the organization for the product?
- Is there a strong economic benefit?
- Is the product a whole product? Does it constitute a complete solution to the need?
- Is the product accepted by trusted suppliers to the organization?

- Will it work within the existing environment (i.e., will there be minimum disruption to existing operations)?
- Does the product meet existing standards promulgated by the IEEE, ANSI, ISO or other standards organizations?
- Have any industry peers adopted or plan to adopt this product?

To summarize, early adopters are willing to take a well-reasoned bet. They are seeking to judiciously balance the risk versus the potential return from adopting the innovation. Inherent conservatism will rule against accepting the levels of risk that a visionary would find normal.

Products and companies often fail to meet the scrutiny of the early adopter when they have not thought through the issues the early adopter is faced with in assuring the successful use of the product in an organization. Besides offering an acceptable risk and return on investment, the product must offer qualities of wholeness.

Case Study
Electronic Data Interchange (EDI)

The Electronic Data Interchange was first used in 1948 with the Berlin Airlift where it was critically important to manage inventories of supplies. Electronic inventory data was sent back and forth via telexes and other communications means to balance what was sent with what was needed in Berlin. After data interchange standards were established in the early 1980s, EDI began to be viewed as a way to enable the efficient exchange of data about sales, inventories, and other commercially important product data. The problem was that while large hub companies such as major retailers and manufacturers could afford the complex and expensive software to create, manage, and send the EDI messages, many of their smaller supply-chain partner (trading partner) companies could not. Therefore EDI never became a whole solution, one that could be broadly embraced and used by all companies in a particular industry. Instead EDI solutions were always partial solutions. In referring to the list of key requirements for an early adopter solution above, EDI fell short in most of the criteria listed.

Unlike the visionary who was willing to contribute money and effort in filling in the holes in the product to fit his needs, the early adopter wants a complete solution.[3] Sometimes the selling company actually can fulfill all the needs of the customer in delivering the whole product. Obviously large engineering contractors like Brown & Root, or large IT companies like IBM, can offer much, if not all, of what is needed. However, even in large companies, the marketer and the engineer must often seek out those business partners and complementary products that give the pragmatic early adopter the sought-after whole-product solution.

In many cases, however, no complete solution ever arrives. The initial enthusiasm that led to successes in the early market does not carry over to the mainstream and no full product or compelling need materializes. Those products that fail to make the leap from the early to the mainstream market are said to fall into the chasm between them, a term originated by Geoffrey Moore, about which more will be said later.[4]

The Late Adopter

If the early adopter was the pragmatist, cautious but willing to take a well-reasoned bet, the Late Adopter is the conservative who wants to wait until the product is proven. To a high-tech salesperson, they are a very tough sell. Whereas early adopters are enthusiastic but cautious, late adopters are only slightly interested in the features of a new device, and only vaguely in the benefits. They want to plug something in and have it work for a specific job. Marketing to these individuals is a new sort of challenge; wherein ease of use, astute packaging, and low cost are key selling points. The product is starting to be viewed as a commodity, a readily available, widely used product. Clearly the PC market achieved this status a number of years ago as systems became available at lower and lower prices and systems were packaged for various markets such as home use, college use, and small business use. This is the portion of the product life cycle where imaginative life extension adjustments make the product easier to buy and use for the late adopter. For example, Microsoft Windows XP, a version that offered a user interface and features that many felt were easier to use than the earlier version, replaced Microsoft Windows 98.

---

[3]Davidow, William H., *Marketing High Technology*. New York: The Free Press (Macmillan), 1986.
[4]Moore, Geoffrey A., *Crossing the Chasm*. New York: HarperCollins, 2002.

The iPod Mini was replaced by the iPod Nano, a smaller, lighter, and more rugged version with essentially the same functionality. Now portable saws and other tools are available with battery power, supplanting the earlier tools that required 110-volt AC power. While the conventionally powered tools are still available, battery-powered tools are now used in many applications where battery power is adequate for the job to be done.

Case Study
The Macintosh: Served a la Mode

> When it was announced in the early 1980s, the Macintosh computer from Apple was a complete integrated system with monitor, computer, and disk drives in one box. The Macintosh had trouble gaining sales in the marketplace for personal computers and for several years languished until it was discovered as the perfect complement to the PageMaker desktop publishing software from Aldus. Until the arrival of PageMaker software in the mid-1980s, the Macintosh met no need as a complete solution for the early adopters and therefore did not gain wide adoption. With PageMaker, early adopters and visionaries found pragmatic business reasons to use the product as a complete desktop publishing system.

The Laggard

Laggards are those people who don't want to have anything to do with technology. To stretch a point, they would still use slide rules if they were available. There are always a few people who want to be left alone, so don't bother with this market segment.

LEAPING OVER THE CHASM

For an engineer involved in new product development, the most interesting stage of any new technology is the early market, or the period of innovation in which the technology is adapted to the needs of the market. In this market stage, innovators will evaluate the product and visionaries will adapt it to the revolutionary needs they envision for it in their businesses.

In the heat of enthusiastic acceptance of the new product, the challenge is to avoid adapting it to too many new applications or markets. The danger is that any small company or department in larger firms will have its hands full with customers all wanting one adaptation or another. Marketing must ascertain which of the probable uses and adaptations offer the most immediate promise of wide-scale use in the mainstream market so that the support and engineering departments will be able to handle demand. Marketing must choose from among the many options available.

- The market that has the largest *economic stake* in the product. This would be the buyer for whom the product can deliver the highest cost savings or monetary gains.
- The market that has the most *immediate need* for the product. This might be a buyer required to use the product to conform to federal legislation.
- The market for which the best *full product solution* can be assembled. This would be a product that brings together formerly separate components in a desirable way.

These are the three main criteria that Geoffrey Moore postulates must be satisfied to establish a beachhead in the mainstream market.[5]

The firms that sit back and allow themselves to be diverted into multiple markets and applications run the risk of falling into Moore's chasm, as described earlier. Many technologies are looking for a way out of the chasm today.

Consider the following.

- artificial intelligence
- neural networks
- pay phones on airplanes
- object-oriented databases

Millions of lines of print found in technical articles, white papers, press releases, and advertising copy have been written about these services and technologies. In spite of the wealth of publicity, these products do not deliver a

---

[5]Moore, Geoffrey A. *Crossing the Chasm*. New York: HarperCollins, 2002.

compelling relative advantage[6] over any capability or product that they might replace. From the perspective of the potential market, they are not worth it.

Although some visionaries may have used the technology to solve some problems in their environment, the pragmatic buyers in the mainstream market have not seen the compelling application that moves them over the edge. While such an application may exist, the pragmatic, conservative buyer in the mainstream market has not been convinced of it.

The focus on building the strength of the relative advantage is crucial in engineering the product. Customers are looking for that major reason to buy that makes all the difference.

Technology often moves in waves. These waves have a wonderful feeling about them, holding out the promise of fame and wealth to all those who climb on and ride. And indeed, technology has delivered fame and wealth for many who were successful with products in this arena. Those who were successful, who figured out how to cross the chasm to the mainstream market, delivered great value for the money by solving a real business problem for a pragmatic, conservative business individual.

The successful technology companies focused on customer needs and constructed a product and accompanying support that met all of the customers' needs. They then successfully differentiated the product so that the customer became aware of just how uniquely suited the product was to solving their problem. With such focused attention, customers bought and used the product.

Skill Builder: Market Research

Discuss with a member of the marketing department in your company how market research is conducted.

Is the marketing department analyzing the past or forecasting the future?

Do they think that analyzing past buyer behavior will help them predict future purchasing?

---

[6]The *relative advantage* of a product over a competitive or existing product solution is probably the most significant factor in determining the success of a product in the marketplace. See Ch. 6 for more information on this topic.

Skill Builder: Life Cycle Stage

What stage of adoption are your products in?

Are they new and being introduced to an unsuspecting market or are they in a later stage of the TALC?

Do you think any of your products are in the chasm?

Discuss with a member of your marketing department how they think these products can be moved out of the chasm.

Skill Builder: Relative Advantage

Do your products have strong relative advantage or are they viewed as middle of the road?

Do customers desire changes to some aspect of your product that could crucially increase your relative advantage over your competition?

5

The Dynamics of Market and Company Growth and Decline

Skill Builders in This Chapter
- Dominant Design
- Moving Forward
- Value Network

Companies must either change and grow or die. The dynamics of modern business demand new and successful products to drive company growth and meet the expectations of customers, investors, and management.

Engineering is the source of innovation that helps meet this business necessity.

This chapter deals with the growth and decline of markets and companies. It discusses the formation and growth of early markets and the effort by the companies in those markets to produce new and better versions of products, fueling the spread of innovation and the onset of competition. All of these efforts are aimed at satisfying or anticipating customer demands. However, significant new studies indicate that when powerful companies focus all their effort on satisfying their current best customers, the opportunity exists for other companies to ultimately steal these customers by being first to address the needs of alternate markets. This is a fascinating new finding that helps in understanding the success of some early-stage companies. This chapter is intended to help engineers and technical professionals understand the larger economic drama they are a part of and their companies' place in a galaxy of competing firms.

GROWTH OF MARKETS—EVOLVING TO THE DOMINANT DESIGN

An innovation can kick off a series of events leading to the formation of many companies, all competing to be the market leader. In each of these companies is a core of persons with a view of the customers and markets that cause their company to take actions leading to the presentation of its best product to the market. From the company insider viewpoint, this can be a confusing scenario as many alternatives for products are considered and some are developed. On the market side of these developments it is not much better, as prospective customers seek to make the best choice from the diverse alternatives available to them in the evolving and changing product landscape.

Consider the development of the auto industry. Figure 5.1 illustrates the growth in the number of firms that entered and exited the industry over the period from 1894 to 1960. This graph shows that 75 firms were in the industry in 1923, the year that Dodge introduced the all-steel enclosed body. This innovation offered increased body rigidity and strength as well as passenger comfort. It also enabled the introduction of the automated stamping of body panels in lieu of the customary hand forming. After its introduction, the all-steel enclosed body became the dominant design for the industry. Two years later 50% of all cars had this type of body, and three years later 80%. Clearly Dodge had taken the industry by storm and competitors moved quickly to adapt to the successful new design. It should also be noted that, after the introduction of the enclosed body dominant design, the number of firms in the industry declined dramatically so that only about 10 firms remained. James Utterbeck points out in his book, *Mastering the Dynamics of Innovation*, that it is difficult to relate the decline in number of firms directly to the introduction of the enclosed steel body.[1] However after the introduction of the steel body in 1923, 13 firms left in 1924 and an additional 13 in 1925. This shift from a hand-formed method to a automated stamping method certainly appears to have had some effect.

As Utterbeck points out, the adoption of a dominant design is characteristic of many industries as they evolve in a competitive marketplace. In fact Utterbeck gives examples from many industries: typewriters, televisions, picture tubes, transistors, integrated circuits, disk drives, and parallel computers among others. The adoption of a dominant design is a signal event for an industry as it will now enable competition in other ways.[2] In the case of the auto industry, this included the development of better components and advanced driving features such as power steering, automatic transmission, and the like. The basis of competition, in effect, shifts from an agreed-upon design to other factors.

Engineers working in a new industry should consider whether or not the industry has achieved a dominant design. Sometimes the achievement is relatively easy to discern, based upon broad industry acceptance of the new design. This certainly happened with the Dodge enclosed body discussed previously, as well as with the IBM PC introduced in the early 1980s. Whereas in the car industry the dominant design ushered in an era of fewer carmakers, in the PC industry it led to the formation of literally hundreds of companies making the new design. However, most failed after a few years and lost out to larger, well-capitalized firms. While clearly a firm should adopt the dominant design if they wish to continue in the industry, it is not clear from history just

---

[1]Utterbeck, James M., *Mastering the Dynamics of Innovation*. Cambridge, MA: Harvard Business School Press, 1994, p. 35-36.
[2]ibid, Utterbeck.

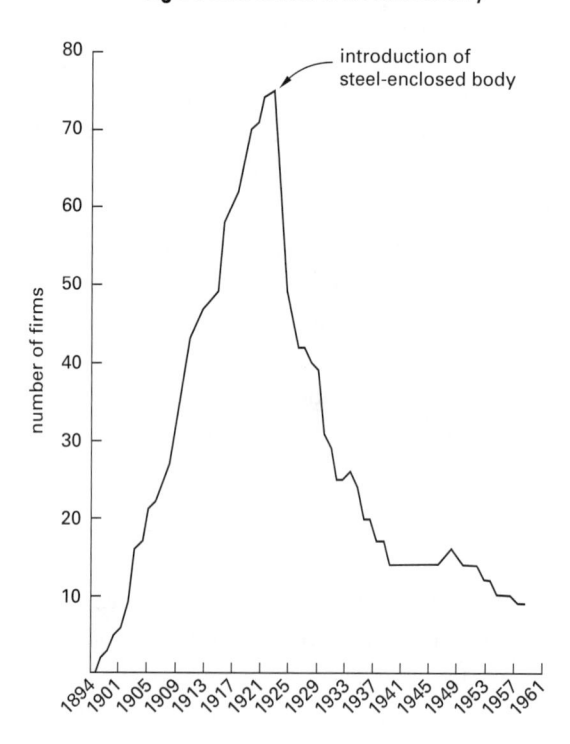

Figure 5.1. Growth of the Auto Industry

[Graph showing number of firms on y-axis (0-80) versus years from 1894 to 1961 on x-axis. The curve rises steeply to a peak of about 75 firms around 1923, labeled "introduction of steel-enclosed body," then declines sharply to about 10 firms by 1961.]

Adapted from Fabris, Richard H., *A Study of Product Innovation in the Automobile Industry during the Period 1919–1962*, University of Illinois (Urbana), PhD Thesis, 1966.

how to survive once the design is adopted. Size, brand name, and capital availability all seem to be factors.

If no dominant design has been achieved, engineering should recognize this fact and remain flexible in its design—recognizing that key ideas and ultimately a dominant design could emerge from any source. In some cases a dominant design may actually be imposed by an outside power, such as the government, as was the case with the 1996 Health Insurance Portability and Accountability Act (HIPAA) transactions rule.

Case Study
HIPAA Transactions Rule

HIPAA imposed a number of new regulations relating to insurance portability and the application of information technology to the healthcare industry. Of particular importance was the imposition of standards for the composition of electronic transactions exchanged between healthcare providers, insurance companies, and clearinghouses. The reason for the new standards was that the industry failed to agree on them on their own, and Congress felt that significant money could be saved through the increased efficiency from using standardized transactions. The new standards were based on the American National Standards Institute (ANSI) X12N (N for insurance) subcommittee's transactions for claims payments eligibility and other healthcare needs. Thus a dominant design was imposed on the industry. Unfortunately, in the rule as written by the Department of Health and Human Services, many loopholes were left in the applications so that there were ways for providers to avoid having to create the new transactions. Rather they used clearinghouses to do this for them—thereby not achieving many of the efficiencies thought to be available. Consequently the impact of a dominant design was watered down and the industry has not benefited from the universal adoption of the standards, although there is slow progress toward that goal. From a marketing view, the market for new methods and the technology for implementing and communicating the transactions never developed to the extent anticipated by many.

The S-Curve—Product Evolution and Displacement

When Thomas Edison invented the electric light, he wanted to displace gaslight. In fact, he planned to use the existing gas pipes as the conduit for running wires to the light bulbs that replaced gas fixtures. The Xerox machine succeeded carbon paper. Refrigerators replaced the need for harvested ice. Calculators replaced slide rules and the personal computer supplanted the typewriter. In each case the newer technology was gradually accepted and replaced the old.

Richard Foster, in his book *Innovation: The Attacker's Advantage*,[3] presents a model for how one technology or product replaces another. Foster's model tracks the inception and steady improvement of a technology as a function of cumulative investment in the technology. The model is shown (Fig. 5.2) as

---

[3]Foster, Richard N., *Innovation: The Attacker's Advantage*. New York: Summit Books, 1986.

an S-curve where the y-axis is used to measure a key performance characteristic of the technology or product, and the x-axis represents total innovation investment.

Figure 5.2. The S-Curve

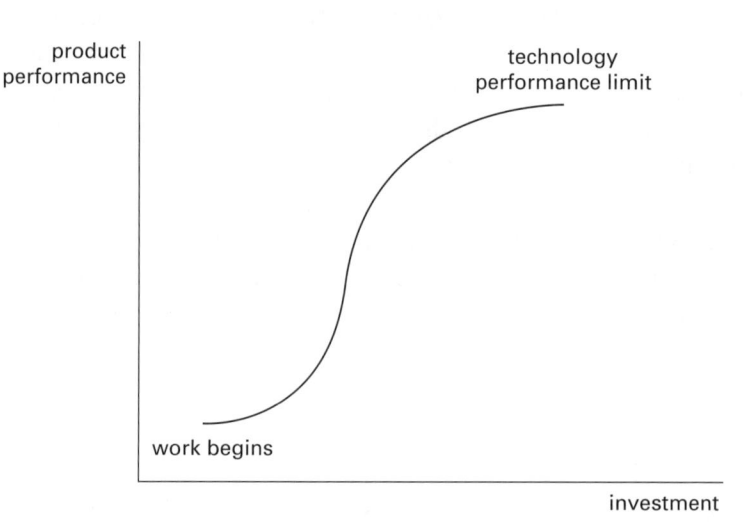

Foster proposes that in the early stages of a new technology or product development, the costs of improvement are relatively high for each increment of additional performance. This is signified by the relatively flat curve early in the product's life. However, at a certain threshold (or inflexion) point, additional high performance is achieved with relatively small investments. The companies that are leaders at this stage stand to gain market dominance by virtue of having accumulated the knowledge to capitalize on their investment. If the technology is successfully deployed into receptive markets at this stage, the leader company reaps the rewards. However, after this rapid growth phase, at a certain point it once again becomes more difficult to extract more performance for the dollar—the S-curve flattens again. The stage is then set for a new competing technology to capture the market. According to Foster, the attacker (usually an upstart company) champions the competing technology. The new company develops technology with performance on a new S-curve, above the existing curve representing the older technology.

5 THE DYNAMICS OF MARKET AND COMPANY GROWTH AND DECLINE 55

Case Study
Dominating the Market

The computer workstation wars of the late 1980s illustrate the S-curve performance/investment concept. Beginning in 1978, when the first VAX computer was introduced, DEC dominated the market for powerful 32-bit minicomputers. The VAX chip and computer systems technology evolved and became the dominant seller in this market, especially for scientific and engineering applications. However, SUN Microsystems Corporation challenged DEC with workstation technology aimed at the scientific and engineering market. They did not enter the evolving minicomputer market. Initially, SUN's workstations were built around the Motorola 68000 chip set. Like the VAX, the Motorola chip used the common chip technology of the time known as complex instruction set computing (CISC). In 1988, however, SUN introduced a new workstation based on a radically new and different technology known as reduced instruction set computing (RISC). SUN claimed that this workstation, the first SPARC, was 10 times as fast as DEC's. On the strength of this new machine, SUN soon dominated the high-performance scientific workstation market. Figure 5.3 indicates the relative performance curves.

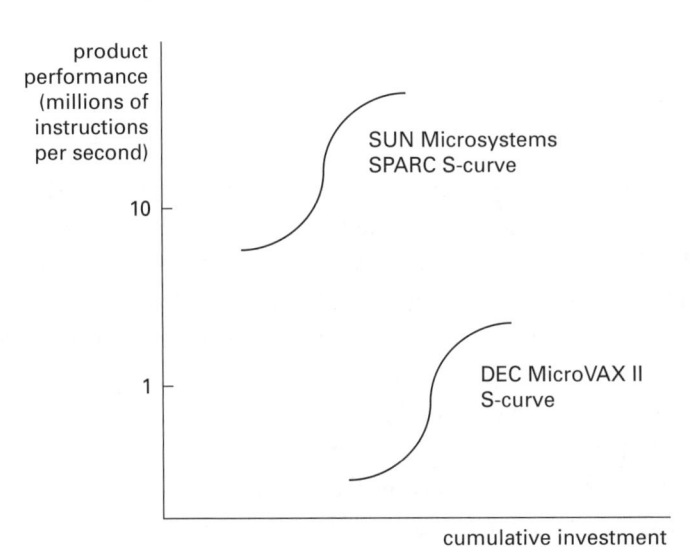

Figure 5.3. The S-Curves of Two Competing Products

PROFESSIONAL PUBLICATIONS, INC.

The SPARC technology, represented by the second S-curve, developed above the curve representing the dominant VAX technology. Its development follows the S-curve parameters until the new technology takes over from the old.

PRODUCT PERFORMANCE

Large performance gains can be extracted from dying technologies even as newer and better technologies supersede them. The reasons for this are unclear, but it is the author's personal belief that the engineers working on the expiring product are spurred on to work harder and extract the absolute most out of the existing technology. Ideas and innovations that would never be tried in a business as usual mode of operation are suddenly found to have merit, and unexpected advances happen as a result. For example, in the space of a few years, an order of magnitude gain in performance was extracted from the VAX chip technology. While this was still less than that available from the RISC technology, it delayed the migration of customers from the VAX, giving DEC more time to introduce the ALPHA AXP chip, its answer to the RISC technology, in 1993.

Case Study
A SPARC That Grew Brighter

> A coterie of DEC's engineers was aware of the threat of RISC technology. A DEC group had experimented with RISC technology in the early 1980s. When SUN announced the SPARC technology, DEC was working on an agreement with a supplier of RISC technology, and in just one year was able to field workstations based on RISC technology that were faster than SUN's. By that time, however, SUN had gained even more momentum, and DEC found that selling the new workstations against SUN was an uphill battle, in spite of better performance and other advantages. SUN had achieved both technical and market dominance in a short period of time, driving a wedge into the market that, by the time DEC had roughly comparable products, was impossible to dislodge.

As will be seen in the next section, another factor operating here was that SUN had begun in a new value network and by the time DEC woke up, SUN was attacking DEC's mainstream customers with its offerings.

As Foster says, products based on technology are always "under attack" because they are subject to replacement by newer alternative technology. Technological change drives the development of new products and the replacement of old products. An engineer who is aware of the encroachment of technological changes can alert his organization soon enough to organize a response. However, in the face of well-organized attacks from a company with superior technology, the astute engineer should assess his chances of survival and bail out if it seems wise to.

What Sustains Companies?

"Will you still need me? Will you still feed me? When I'm sixty-four?"[4]

Everyone knows that customers sustain companies…right? Customers buy the products, provide valuable insight into what is needed next, allow the marketing guys to buy them lunch…what could be more sustaining than a company's customers? When we get to be sixty-four we want to be making and selling products to the same customers we had back when we were young.

However, in some cases listening to a company's customers is like contracting the flu. It might lead to a prolonged period of illness and could be fatal. Professor Clayton Christiansen of Harvard Business School believes this is the case in circumstances that he makes clear in two insightful books: *The Innovator's Dilemma*[5] and its sequel, *The Innovator's Solution*[6]. Here Christensen lays out radical ideas that he backs up with case after case of real companies that succumb to defeat at the very hands of their customers. As a consequence of focusing on serving their existing customers (and making good money by doing so), they ignore what is going on outside their immediate sphere of interest.

---

[4]Lennon, John, and McCartney, Paul. "When I'm Sixty-Four," sung and recorded by the Beatles, 1969.
[5]Christensen, Clayton M., *The Innovator's Dilemma*. New York: Harper Business Essential, 2003.
[6]Christensen, Clayton M., and Raynor, Michael E., *The Innovator's Solution*. Cambridge, MA: Harvard Business School Press, 2003.

This immediate sphere of interest is the company's value network, the "money pot" that enables a group of companies to thrive and prosper in serving an end customer, while trading products and services among themselves to meet the needs of the end customer.

What is a value network? It is the collection of companies that make a product for an end user. As an example, Christensen uses the disk-drive industry. From the standpoint of researching business practices, this industry provides a valuable characteristic: it regenerates itself and evolves very rapidly. Over the years many generations of disk drives, each with steadily advancing improvements, have been developed and marketed. The disk drive industry is like the common housefly used for genetic research—both reproduce very rapidly. As with housefly generations, a researcher can collect data on the success and failure of generations of drive products because of the rapidity of change and population growth.

Within the disk drive industry, the value network is composed of all the companies that make the components and final drives for the end user. These companies include the read/write head makers, the motor makers, the power supply makers, and those that make advanced chips used in the control electronics. The value network also includes the company that integrates the drive with the rest of the computer components to make a final assembled computer.

The significance is that the composition of the value network and its profit margin can change dramatically from one end user application (that supports the network) to another. For example, the 5.25 inch disk drives used in early minicomputers sustained a vastly different value network from the smaller drives used later in laptop computers. There was a certain profit margin expected by all the members of the 5.25 inch value network. When the 5.25 inch drive makers developed and evaluated the 3.5 inch drive applications, they said, "No thanks! Our customers don't want these, and we can't make any money on them anyway."

In fact, the appropriate application for smaller drives was just emerging in a separate market, the laptop market. The 5.25 inch drive users, minicomputer makers, were busy at work making and delivering minicomputers. In order for mini makers to succeed in laptops, they had to develop a new value network of suppliers who could operate effectively with the alternate performance

5 THE DYNAMICS OF MARKET AND COMPANY GROWTH AND DECLINE

characteristics (low weight, smaller size, and restricted power consumption) that were required in the laptop market.

This migration disrupted the existing value network. New companies that made the 3.5 inch drives were able to find a new and early market with just-emerging laptop companies that were developing and operating in a totally new value network. Later, as the 3.5 inch technology improved, they sold the 3.5 inch technology into the 5.25 inch mini market, displacing or disrupting the existing minicomputer disk drive value network. They were now able to deliver superior performance to the 5.25 inch driver makers at a lower cost.

What happened is that the 5.25 inch drive makers satisfied their existing customer base but, due to their dependence on an existing value network, were unable to respond to the encroaching threat from 3.5 inch drives that began in a new market, laptops. This scenario is illustrated in Fig. 5.4.

In another example, Christensen discusses how the backhoe market developed shortly after World War II. During the pre-WWII era there were about 30 established manufacturers of excavation equipment, all of which used the dominant technology of that time, cable-actuated and lifted excavation buckets. Just after WWII it became apparent that hydraulically actuated equipment

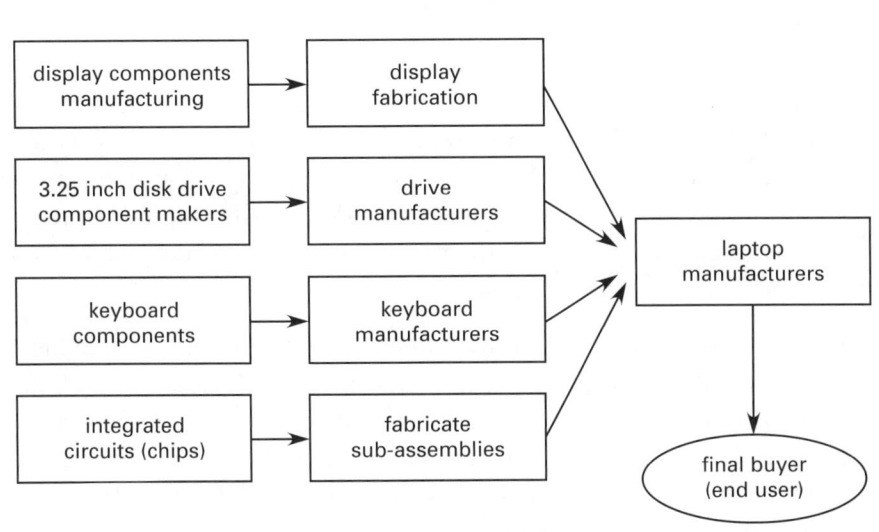

Figure 5.4. Value Network[a]

[a]Not shown are sales organizations and distributors that could also be part of the value chain delivering products to end-users.

PROFESSIONAL PUBLICATIONS, INC.

was feasible for the excavation market. However, these hydraulic mechanisms could only dig and carry loads smaller than what existing products were capable of. While some existing manufacturers developed hydraulic equipment, they found that their existing customers were not a good market because the equipment did not meet their needs for dig and load capacity.

Meanwhile, companies outside the mainstream developed hydraulically operated backhoes, so called because they were attached to the back of a truck or tractor. Early backhoe manufactures marketed shovel width and truck/backhoe maneuverability and found a ready market for light jobs distinct from the uses of the dominant cable-actuated equipment that served a market requiring equipment of much higher capacity. Over time, hydraulic weight and strength limitations were overcome, and hydraulic capability displaced or disrupted the cable-actuated earth-moving equipment.

As a result, of the 30 firms dominant in the 1950s that made and sold cable-actuated equipment, only four survived into the 1970s. These are household names today and include such firms as International Harvester, Caterpillar, and John Deere.

In his books Christensen gives many examples of companies run by very smart and capable management that were unable to respond to and successfully counter the threat posed by companies that started by applying technologies to a market not seen or unable to be served by the existing suppliers and by creating an alternate value network.

As seen earlier in this chapter the progress of a particular technology can often be described by an S-curve representing performance capability (y-axis) against investment (x-axis). However, new technologies almost invariably arrive to compete with an existing capability causing a shift of products to new ones based on the technology in the new S-curve. Using the new S-curve is, in Foster's view, the attacker's advantage.

Christensen shows that actual growth and decline of new products and industries is subtler than just jumping onto a new S-curve. His research postulates the existence of two distinct types of technology: sustaining and disruptive. Companies focus on their sustaining technologies that are bought by their existing, sustaining customers. These companies are doing what marketing tells them to do—satisfy their existing customers. However, in plain view of all, the seeds of disruption are sown because, while they are satisfying their

5 THE DYNAMICS OF MARKET AND COMPANY GROWTH AND DECLINE

existing customers, other companies and their engineers are hard at work developing new technologies and products for a new value chain. In time this new upstart company, with its roots in a totally new value chain, can disrupt the existing companies by upgrading products so that they are of superior value for the money in the old markets. For example, in time the 3.5 inch drive became the industry standard drive for minicomputers and the laptops.

The dynamics of technology evolution, industry and market formation, and product development presented here are complex. Even with this brief overview, though, key ideas are presented that can help an engineer understand in broad terms the marketing and economic milieu. Moreover, it provides a historical context to show that this type of change is not new and that much has been learned from experience. Although the experience is in some ways frightening, in that it catalogs the disruption and decline of many firms, at least one can appreciate that this does happen and take steps where possible to avoid past mistakes.

Skill Builder: Dominant Design

Is there a dominant design for your product set, or are a number of firms battling to establish it?

What do you think are the key characteristics that will affect the eventual development of the dominant design, or if one exists already, what key characteristics comprise it today?

Skill Builder: Moving Forward

What technology or product set is being shoved aside by the new developments in your industry? Where do the new developments sit on the S-curve?

Skill Builder: Value Network

Who are the key members of your value network?

Who besides your company offers the most value to the ultimate end consumer?

How could you company add more value?

6

Engineering Insanely Great Products

Skill Builders in This Chapter
- Using Quality Function Deployment
- Trading Off
- Salability

What constitutes a great product? Or, as Steve Jobs would say, an insanely great product? (This was his proud term for the products of his company, Apple Computer.) Take a look at the product that put Apple on the map, the Apple II, a home computer delivered in 1978 at the very birth of the microcomputer era. An Apple II purchased in 1978 came with a wealth of features that made it extremely usable and adaptable to many applications.

For those who don't recall the Apple II, it is worthwhile reviewing the array of features that this product delivered for about $1200 in 1978. Built on the 8-bit 6502 microprocessor (similar to the Motorola 6800), the Apple II had a built-in BASIC interpreter as well as an assembler for programs written in the 6502's assembly language. The Apple's software was contained in PROMs (Programmable Read Only Memory) on the motherboard, but extra sockets were provided to contain custom programs that could be triggered from other programs. In fact, the code for the Apple's operating system was made available and could be accessed and changed. Therefore, the soft-wired features could be changed and a PROM reblasted containing the new functionality.

There were seven easily accessible bus slots so that custom circuit boards could be built to add new communications, graphics, sound, or other features to the system. Visual output could be displayed on either a television set (using an RF modulator) or on a monitor. The complete system was housed in a hard plastic case that provided both durability and easy access to the components. It was used in widely varying adaptations for many environments, from elementary schools to the Department of Defense.

The Apple II sold and sold and sold. It was a huge overnight success.

This chapter deals with some of the marketing considerations that go into delivering products like the Apple II. As anyone who's been involved in an engineering project knows, it comes down to questions of quality, time, and money. Many times the project comes down to which one is given up to ensure the other two. But customers want high-quality, inexpensive products and they want them now. This chapter addresses how to make judgments about these three factors—quality, development time, and productivity (money)—while still providing what the customer wants.

QUALITY

Quality is much more than building products that won't break. Quality is an abstract idea that becomes concrete with products like the Apple II, which was not only durable but satisfied a wide spectrum of customer expectations and needs—needs they probably didn't know they had at the time of purchase.

Companies often use attention-getting expressions like "Quality Is King" or "Quality Is Job One" to help customers and employees recognize the importance the company attributes to quality. Unfortunately these types of expressions sometimes do little to impress either the customers or the employees, but they may make the executives feel like they've earned a day's pay. The distinction comes if the company makes a commitment at all levels to following through on the concept, with programs and products that deliver the quality that customers want. What's needed is recognition that quality in business and engineering is a broad concept that covers all aspects of a product—concept, design, engineering, manufacturing and marketing stages. In other words, a brilliant product concept can result in a flawed product if any of a host of other activities do not meet the mark.

Sidebar 6.1: Achieving Zero Defects

When *zero defects* became a business fad, the idea was that product quality would improve by calling attention to problems through this campaign with a catchy title. Unfortunately, the campaign drew attention to defects after they had happened; little was done to prevent the defects from going into the products in the first place.

Later on manufacturers abandoned the idea of zero defects and switched the focus to the manufacturing process itself. By reducing the amount of variation during the manufacturing process, they sought to reduce the defect levels within those products. One such process, called the Six Sigma method, could potentially reduce defects to 3.4 defects per million opportunities. The Six Sigma methodology focused on understanding variation in processes that produced the critical Xs, which were the variables that affected the major quality metric, popularly known as the Voice of the Customer. While the Six Sigma level is extremely difficult to attain, the focus on processes can lead to improvements that result in better levels of quality. With success in application to manufacturing processes, the method was later applied to financial services, marketing operations, and many other aspects of business.

In order to satisy customer needs, all of these components in the development and marketing process must be present if they are to contribute to quality. Suppose a product is well manufactured and operates extremely well, but its physical appearance doesn't win over its intended customers. In this situation, the result would be the same as if the quality were poor; it wouldn't sell. Why? Simply stated, because it didn't satisfy all of the customer needs.

Case Study
Another Success at Apple

> Another of Apple's successes is the iPod, a paramount example of detailed attention to all phases of the product development process. With the iPod, Apple achieved stunning success with an incredibly great product that combined beautiful and intelligent design with rugged and innovative engineering. With iTunes, its music download service, Apple delivered a complete product that has won the hearts and ears of millions.

In a study conducted by the Marketing Science Institute, called the PIMS study,[1] researchers tracked product success experienced by many companies over an extended period of time. One of its most important findings is that product quality is the highest determinant of long-term market share in any industry, underscoring the importance of focusing on quality at all stages of a product's life cycle.

What does this statement really mean? Is it saying that in the car market, for example, because Mercedes is of higher quality than Chevrolet, Mercedes will sell more cars than Chevrolet? Clearly not. The missing piece is market segmentation. The statement means that in each company's respective market segment, quality will win out over time. In the car market, a major factor that determines segment is price.

Mercedes sells to those who are willing to spend more than $30,000 on a car, the high-priced segment. Chevrolet targets the mid-range market segment, those customers who want to spend less than $30,000 on a car.

---

[1]Buzzel, Robert D., and Gale, Bradley T., *The PIMS (Profit Impact of Market Strategy) Principles*. New York: The Free Press (Macmillan), 1987.

Without faulting Chevrolet, Mercedes should be able to build a higher quality car than Chevrolet if only because its greater price helps finance the costs of engineering development.

Chevrolet's logical goal would be to build the highest quality cars in its targeted market segment. However, when Ford introduced the Taurus, Chevrolet's dominant market share in the mid-range segment quickly succumbed to the higher quality evident in the Ford product. Over time, Ford steadily increased its market share in the mid-range segment as a result of its stunning success in delivering a markedly higher quality product than Chevrolet.

Mercedes experienced a similar situation when Lexus entered the high-priced market segment with products many customers thought of as being superior to Mercedes'. Mercedes quickly and drastically lost market share in the high-priced segment. In a short time Lexus built tremendous sales momentum in the luxury segment, so that in 2005 Lexus outsold not only Mercedes, but also BMW, Cadillac and all other luxury brands. All of these brands have been in business far longer than Lexus.[2]

Building in Quality

Much has been written about how Japanese industry set a quality standard for the world. Building upon the ideas of two Americans, W. Edwards Deming and J. M. Juran, the Japanese instituted methods and procedures to continually improve products. These techniques led to stunning product success in industry after industry, market after market. They woke the world to what was achievable by continually improving quality. Overcoming the destruction caused during World War II, from their relatively small island country the Japanese seem to have used quality improvement as their strategic weapon in their world domination of a number of industries. On the other hand, as mentioned at the outset of this book, for American manufacturing the years after World War II were years of robust growth as the American economy prospered and consumers bought whatever could be obtained. Yet barely 20 years after the end of the War, Japanese car makers entered the American markets and began their campaign to use quality as a means of gaining market share. They

---

[2]*Barron's*, "Lexus' Next Act," September 26, 2005, p. 41.

were greatly helped in this endeavor by the complacency of American car makers, which continued to create and sell products of relatively low quality.

Evolving from the work of Deming, Juran, and others, an important idea called Total Quality Management (TQM) has emerged. TQM seeks to instill a universal view of quality within an organization, focusing on how everyone, from the president to the rank-and-file, make active contributions to the company's continuous striving for quality

A systematic way of achieving product quality—part of the TQM philosophy—is found in the method called Quality Function Deployment (QFD). QFD focuses on achieving customer satisfaction by analyzing customer needs, competitive influences, and company capabilities. Customer needs (voice of the customer) and other factors are modeled in Table 6.1, called the House of Quality. Taken from the house-building process, the factors that are judged to be important are assigned numerical values and weights dependent on their relative importance. Through this model the most important customer needs and design issues emerge and evolve from discussions and reflection. The house provides a way to bring all factors together so that they can be considered as a whole—much in the same way an architect brings together various design elements and systems.

In Table 6.1, the voice of the customer is represented by the needs expressed in the left-hand column. Current product features are listed in columns across the top (proposed features might also be included). S, M, and W indicate the correlation between the features and customer needs. (The S stands for strong correlation, the M for medium, and the W for weak.) Here the proposed product scores 25 out of a total of 33.

More useful than the total score is that, with the QFD process, all customer concerns can be considered in an objective matrix framework. In making design and engineering trade-offs, this use of reasonably objective measurements helps reduce arguments based solely on opinion.

An example of this occurred at Hewlett-Packard with an engineering work-effort tracking and reporting system.[3] The existing system was not meeting users' needs, so HP designed a QFD matrix to improve it. The evaluation produced a score of 12 out of 50, which confirmed that the existing system was

---

[3]Grady, Robert B., *Practical Software Metrics for Project Management and Process Improvement*. Englewood Cliffs, NJ: Prentice-Hall, 1992, pp 33–34.

6 ENGINEERING INSANELY GREAT PRODUCTS

Table 6.1. An Example of a House of Quality

voice of the customer \ product features	single board	DMA interface	multiple I/O	dual processor	resolution choice	multiple OS support	VT emulation	memory expansion	customer value	system rating
standard terminal emulation							S		3	3
multiple operating system support						S			3	3
fast graphics drawing speeds			M		M			M	5	4
multiple display support				W				M	4	3
low cost	S								4	3
minimum space occupancy	S								5	4
rock steady pictures				W					4	1
choice of resolutions					S				3	3
worldwide support		W							2	1
								total	33	25

key
S = strong relationship
M = medium strong relationship
W = weak relationship

not meeting customer needs. Improvements were discussed, and those that engineers felt were needed were placed within the QFD framework and scores assigned. With the improvements, the ratings increased to 44 out of 50. Users were reportedly so unhappy with the existing system that it was thrown out instead of being used as the basis for the new system.

Successful products are not built without attention to detail. Engineers, however, focus on vast amounts of detail specific to their engineering specialties. A House of Quality matrix helps to organize the entire spectrum of considerations and details into a unified whole that centers on the customer. It provides a way of quantifying and prioritizing trade-offs among many variables to achieve the maximum overall product quality. Through visualizing product quality this way, a higher level of overall customer satisfaction can be achieved.

PROFESSIONAL PUBLICATIONS, INC.

DEVELOPMENT TIME AND PRODUCTIVITY

Besides quality, the other two factors critical in building successful products are development time and productivity. First, a company would not succeed if it delivered products that were late to market. Second, the products must be priced right—not too high for people to buy nor too low to make a profit.

Development Time

Development time is the time from a product's inception to the time of first revenue ship (FRS), which means that the product has been shipped to its first customer and the customer is expected to pay. Products can be on time, early, or late to market. Those that are early may fail in the market because there is no supporting market infrastructure that has prepared the market for the product (*see* Ch. 8). Late products may miss the market window of opportunity. In missing the window, the product may be one among many and it may be difficult to displace others that are already in the market.

Productivity

Productivity is a measure of how clever engineers are in developing products that deliver maximum value for the money, time, and materials (i.e., resources) required to develop the product. An equation that illustrates this relation is

$$\text{productivity} = \frac{\text{value delivered to customer}}{\text{resources employed}}$$

The numerator of the productivity equation takes many factors into account. As a simple example, suppose that a company were able to produce a jet airplane for sale to commercial aircraft markets for $1 million per plane. It could be said that the value placed by the customer on this plane was $1 million. The marketing department estimated that over the lifetime of this plane 200 could be sold, so that the total value delivered to customers worldwide is $200 million. Most customers will perform some analysis on the value of the plane to them using such metrics as Return on Investment (ROI), for example. They will attempt to verify that their ROI from purchasing the plane is, say, above 12%, a metric that they use to justify investments.

Mack Hanan, in his book on market penetration, says that a business customer will make a purchase decision based on one of two factors. The purchase will make money (improve revenue in the company) or save money (lower costs).[4] In this example, the company could have made various assumptions that by being able to attend meetings in foreign countries with their new jet plane, they would be in a better position to bid on contracts (make money). They also could have factored in cost savings in executive time by not spending time in airports, not following commercial aircraft schedules and the like. In any case the combination of all these factors, in the customer's view, added up to dollars saved or dollars earned.

The net result of all these calculations from the customer's view was that the plane had a value of at least $1 million.

In other types of markets, such as consumer markets, decisions are often made on the basis of many factors besides cost savings and income earned. For example, decisions may be made for emotional reasons. In these markets, it is often difficult to find a quantitative measure for value. Over time, a product's value is determined by what customers are willing to spend across the sales (usually) of many units. But when the product has not yet been introduced to the market, it can be hard to forecast what customers will be willing to spend.

Value can also be hard to quantify in government markets. Competition is not always a factor in government procurements. Even where it is a factor, there may be only two competitors. In such cases, the government often awards contracts to both in order to keep both in business. Clearly the government finds value in maintaining both companies, but it's hard to measure this value and weigh it against other values.

In considering the denominator (resources employed) of the productivity equation, the accounting department can usually supply the information needed. Accounting keeps track of what has been spent on product development, selling effort, manufacturing, and the myriad other expenses that go into

---

[4]Hanan, Mack, *Successful Market Penetration. How to Shorten the Sales Cycle by Making the First Sale the First Time.* New York: AMACOM, 1987, p. 4.

bringing a product to market. Higher expenses mean lower productivity for the product—an inverse relationship within the productivity equation.

Until a product's productivity has been compared with other past and present company projects, the team has no way of knowing whether the product achieved high or low productivity. It is a metric without a basis for comparison.

One common rule of thumb is that productivity should be 20, meaning that a product should return 20 times in revenue what it cost to build. The problem with this approach is that revenue is a "builder-centric" measure and not a "customer-centric" one. However it is implicit that customers have made their own calculations of value based on what the product will cost, and have determined that the product's value is adequate for the price set.

The astute marketer will focus on determining just how much value is delivered to the customer, as this is what determines the success of the product. An estimate of this value can be used to set the price, and as more is known about the costs and metrics used in the customer's business, a more accurate assessment of value to the customer can be made.

THE PRODUCT LIFE CYCLE

The product life cycle, as shown in Fig. 6.1, depicts products as dynamic entities with a fixed life span, much like people. Development time and productivity figure prominently in determining the revenue achieved from the product, which the model plots over time. As can be seen from the figure, extending the development time will mean that revenue is not achieved until later in the product's life. While it is possible that the amount of revenue achieved may be the same or even more, from the point of view of any project planner, later revenue is more uncertain.

On the other hand, the productivity built into the product will help in extending the product's lifetime, thereby increasing revenue. The ability to achieve product variations (flexibility) and ease of product extension can deliver product enhancements that extend its life and increase its revenue.

6 ENGINEERING INSANELY GREAT PRODUCTS

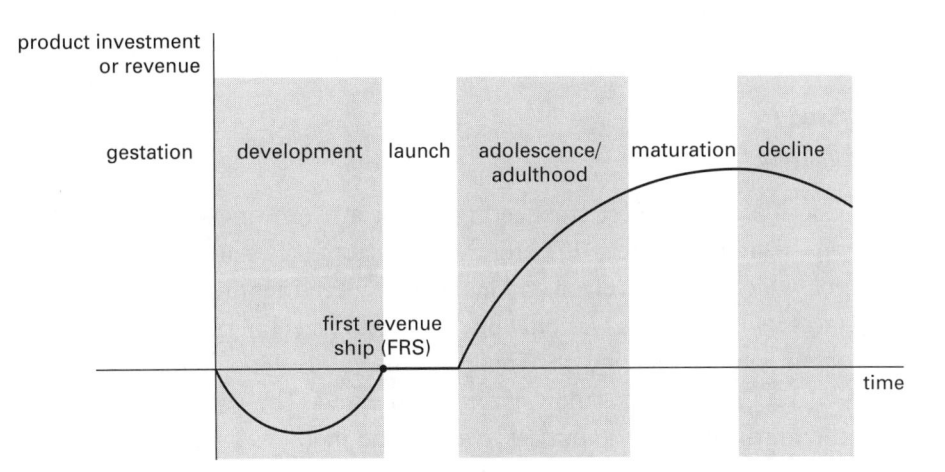

Figure 6.1. The Product Life Cycle

[Chart showing product investment or revenue vs. time, with stages labeled: gestation, development, launch, adolescence/adulthood, maturation, decline. The curve dips below zero during development, crosses zero at "first revenue ship (FRS)", rises to a peak during maturation, and declines afterward.]

Unlike a static view of products, the product life cycle cultivates the idea that products have stages. These stages include periods of gestation, development, adolescence, adulthood, maturation, and decline. This life cycle curve helps project the impact of slips in the product schedule and the amount of revenue expected at any stage. As the product's life actually proceeds, developers can assess the impact of deviations from the curve and take any actions necessary. The project formally starts at time $t = 0$ with a budget, a schedule, and engineering objectives.

TRADE-OFFS AMONG QUALITY, DEVELOPMENT TIME, AND PRODUCTIVITY

The previous sections discussed the three key variables for product success: quality, development time, and productivity. Initial project agreements contain specifications for

- features
- timeline
- budget

It does not take long once a project starts for these to be traded off against one another. This may be the result of issues that arise after the planning has been done, unforeseen events, a desire to push on with a project even with evidence that many factors have not been planned for, or the realization that the project has poor prospects.

Case Study
Value Delivered

Consider the greatest engineering feat of modern times—man's landing on the moon in July 1969 and the astronauts' safe return. For this project, which was conceived and launched by President Kennedy and Congress in 1961, there was only one acceptable level of quality: landing on the moon and safe return. There was also a deadline: before the end of the 1960s. So two of the variables were fixed. The goals were met, but the third variable, the budget, reached the moon as well. Who were the customers for this project? The 200,000,000 or so U.S. taxpayers. No one who witnessed the moonwalk of 1969 via television can ever forget that event. Was customer satisfaction achieved? Absolutely!

If a good understanding of quality as it pertains to the product under development has been achieved through a process like QFD, then a reasoned basis for making trade-offs among various key quality attributes will exist. In other words, certain quality attributes of less importance may be dropped. While it is not desirable to drop any quality attributes, it may be necessary if development time or budget is to be maintained. Another option may be to appeal for an increase in budget for the project. An increase in budget may stave off lowering quality or increasing development time. While it is difficult to generalize, the information shown in Table 6.2 may be helpful in considering trade-offs.

Table 6.2. The Effects of Modifying Development Time, Budget, or Quality

action in product development	result
development time is increased	increased costs (more payroll) lower revenue (less selling time)
budget is increased (e.g., to keep quality at desirable levels)	increased costs
quality is lowered (features removed)	lower revenue due to less desirable product

Note that, of the three actions in Table 6.2, increasing development time is the only one that both increases costs and lowers revenue. The importance of this effect cannot be overemphasized—project schedule is crucial because the effects are amplified across both costs and revenue. Another undesirable effect is the possible loss of the market window, mentioned earlier, which may lead to a precipitous decline in market leadership.

Increasing the budget by putting more people on the team or buying labor-saving equipment will affect payback. On-time product sales can deliver revenue at a considerable multiple over amortized product development costs, so increasing the budget to maintain the schedule and begin actual sales as quickly as possible makes sense.

By choosing to lower quality, the project team confronts lower revenue due to unsatisfied customer needs. The product will not sell as well over its entire product life. However, if quality goals are maintained, the chance of building a winning product increases, which increases the probability of delivering a satisfactory return on investment (even if the investment was increased to maintain quality). Thus, even from a practical project management standpoint, quality should be preserved.

Cool Judgment in Confusion

It is not usually possible to make such trade-offs among quality, cost and development time in an atmosphere of due consideration. The development staff and management are just too harried and pressed to make these decisions with cool detachment. Often, quality features are thrown out along with the development time and the budget. In these situations, investing time in developing a House of Quality matrix will pay dividends by providing a background to help analyze the impact of time slips, increased spending, and foregone quality dimensions (features) on the overall success of the product.

Referring back to the product life cycle in Fig. 6.1, the growth phase begins with the product launch. In this stage, marketing figures prominently in finding the customer and telling him what the product will be good for and how it will be used. If the product is an innovation, the customers are the technology innovators and visionaries, corresponding to the first two stages of the technology adoption life cycle (see Ch. 4).

Engineering is also frequently called upon to explain the technical features of the product to the early market and to address customer concerns. They may be requested to adapt the product, or to add complementary features to the product to more fully address the needs of the chosen market or an alternate market. While the company and the engineering staff may be delighted to have the customer attention, the requested changes can affect how well the product meets the needs of the originally proposed customer base or market. Any proposed engineering effort, then, that diverts product development from its original target should be carefully scrutinized by both marketing and engineering to understand the impact of the new development on the original targets. Questions such as these should be asked: "Why haven't we achieved forecasted sales to our original target market?" and "What changes have occurred in the original plans?"

Case Study
Changes in Customer Needs

In a computer graphics circuit board development, two major shifts in customer focus occurred during the gestation and development time for the product. The first was the encroachment of the personal computer (PC) on traditional markets for graphic display. The PC was an adaptable, low-cost platform that met needs of certain industrial display markets. The second was new customer interest in standards-based software, which was used to develop graphics displays. A standard called graphical kernel system (GKS) was gaining popularity and would enable a user the freedom to move his program (written with GKS) to a variety of hardware display environments, rather than being locked in to the language and the hardware of the hardware manufacturer. As development of the system was concluded, it was apparent that the changes definitely affected the markets for the display system. Providing these additional capabilities meant hefty increases in the engineering budget and therefore an extended and more risky payout for the product.

These were rude shocks to confront. The additional funding required was not received. However, the flexibility and adaptability built into the circuit-board design uncovered new and unintended markets in applications that had not initially been considered. In short, the early market period was a time of considerable seat-of-the-pants adaptation.

Just as with people, product transition from the adolescent to the adult stage is sometimes anything but smooth. Sales may not develop as quickly as expected. There may be nagging technical problems. Customers may need more technical support than allowed for in the initial plan. As seen in Ch. 4, a clear direction and focus for marketing efforts helps in this transition, so that engineering and marketing are well in tune to succeed in a chosen market.

Finally, a product that moves into the maturity stage must fit well into markets that are characterized by people who want very low risk. If the product is still being sold at this stage, the mainstream market community has accepted it, and even late adopters will buy it.

Unlike people, products are amenable to product-life extensions. Boosting performance or adding features are two ways of reviving flagging products. These product variations are often successful because the product's marketplace has been established and there is high product acceptance with existing customers. Therefore, product-life extensions can be used to squeeze more revenue out of products on the decline. With software products, new versions are seen as both a means of correcting many of the bugs of past releases and of adding considerable capability in response to customer requests for improvements. The adaptability and flexibility that has been built into the product at the start can be exploited even more at this stage.

Sidebar 6.2: Ship Regardless

"The alternative to shipping products with small flaws is to fix every bug and miss out in a competitive market," says Bill Goodhew, the former product manager at Lotus Development Corp. and current president of PeachTree. "If every software vendor tried to sit there and ship out bug-free products, we would never have any software." Other vendors agree that they can't afford to put out a bug-free product. "PeachTree would rather use developer's time to add new functions to products than to fix every minor glitch," he says.[5]

---

[5]*The Wall Street Journal*, January 18, 1995, p. B1.

FIVE FACTORS TO GIVE YOUR PRODUCTS A JUMP START

How can a product's reception in the market be improved? Are there marketing factors that help boost the probability of a satisfactory reception by the market? To start with, if the product meets its quality goal, then most of the product requirements that were established at the product's outset have been met. There are five additional factors that will help ensure a successful product launch.

These other five factors are from a field of research known as diffusion of innovations, begun by Everett Rogers in the 1930s.[6] As discussed in Ch. 4, Rogers was interested in helping farmers adopt innovations in crop production that would increase yields and improve crop quality. Working with farmers in the Midwest, Rogers found that the successful adoption of these innovations in a market community was best achieved with attention to the five success factors listed in Table 6.3.

Prospective buyers will compare most innovations to whatever it is they are replacing. As a result, a customer goes through an evaluation process focusing

Table 6.3. The Five Innovation Diffusion Factors

success factor	explanation
relative advantage	How much better is this product than the one it will replace or its closest competitor?
trialability	Can a potential customer use the product in a trial mode? Is it relatively easy to try it out?
observability	Can the product be observed in action without significant commitment or resources to get it working?
compatibility	Does this product work with other parts of the operating environment or overall system?
relative complexity	How hard is it to understand? (Products that are understood are easier to market.)

---

[6]Rogers, Everett M., *Diffusion of Innovations*. New York: The Free Press (Macmillan), 2003.

on these five factors. Most important is the relative advantage of the product over what it will replace. The prospective customer judges the new capability in terms of how much better it will be for the desired specific task.

Only a significant advantage justifies a switch. A rule of thumb is that the new capability must have a performance advantage of ten to one over the old product. If the product is an adaptation, then relative advantage, while still important, will not be as crucial, because the risk of adoption by the customer is less.

Lessen the customer's perception of risk by focusing on how the customer will actually try out the product and whether it can be observed in operation at another site. Evaluation copies, demos, and samples are key in establishing customer confidence in new products. Equally important would be product adaptation by other individuals in their market community. The observability factor brings to the fore the importance of customer references in a particular market community.

If a product passes these hurdles, the customer will want to know how compatible the product is with his existing operation. Thus, even though there may be numerous computer languages newer and better than COBOL, an MIS director will not switch if the costs of training and program conversion are significant. In a very different market a customer will ask: "Will the new pump physically fit into the space that I have available when I have to run the old one in parallel with it during a trial period?"

Finally, a buyer will evaluate how complex the product is. Products that are more difficult to understand than an existing product and that cannot be easily explained have a tough time in the market.

> **Skill Builder: Using Quality Function Deployment**
> Develop a House of Quality for a product that you are working on. Make a list of the product's important features. Then make a list of the important functions that you think customers want in the product. Arrange them in a matrix similar to that in Fig. 6.1. Rank the features against the customer functions desired using the S, M, and W rating scale.
>
> How close are the product's features to meeting the functionality desired?

Skill Builder: Trading Off

What would you give up in a project you are working on: development time, quality or budget?

What product features would you give up in order to make your original or current schedule?

How much of an impact on the product sales do you think this will have?

Skill Builder: Salability

Consider the five innovation diffusion factors in Table 6.3.

How well do you think you are accomplishing these factors in your product development?

Discuss this table with your marketing team and see how they can help with achieving more emphasis on these factors.

7

Managing Risk in Projects

Skill Builders in This Chapter
- Assessing Risk in a Current Project
- Specification Review
- Understanding Your Market Risk

"You gotta know when to hold 'em, and know when to fold 'em"[1]

Poker is an apt example for this chapter's discussion. Take the game called five-card draw. Here each player is dealt five cards. After reviewing their hands, players have the opportunity to bet or fold. Players who haven't folded can either keep the cards dealt or discard as many as needed in the hope of strengthening their hands. Players can then fold or bet again—thereby showing their relative confidence in their hand. They're betting on the strength of their hand relative to the competition. Is it better or worse than the hands of the other players?

Product development projects can be thought of in the same way. Call the first round of betting the initial market assessment for a proposed or existing product. Think of the subsequent rounds of betting as the continuing assessment of the project, taking into account all the factors that may impede a successful launch. At each stage of the project, risk is assessed and a new "bet" made in terms of the expenditures of funds and other resources. Engineering product or service projects can be assessed as having two components of risk: market risk and technical risk. The dividing line between the two is the product specification, a marketing department's written assessment of what is needed to win the market.

This chapter focuses on a better understanding of risk, and of ways to manage it so that more projects beat the odds, thereby meeting their marketing and technical goals.

PROJECTING PROJECT SUCCESS

Market Risk (MR) is the probability that the product will not meet the marketing goals set for it in terms of market share, price attainment, volume, and profit. *Technical Risk* (TR) is the probability that the product will either not be delivered or meet performance goals to satisfy the market needs. Both can be expressed as probabilities between the numbers zero and one. A product deemed to have an MR factor of 1.0 means that there is no chance that the product will succeed in the market. A product with an MR factor of 0.1 means

---

[1] Schlitz, Don, "The Gambler" sung and recorded by Kenny Rogers, Hal Leonard publisher, 1978.

7 MANAGING RISK IN PROJECTS

that the product has an extremely good chance (90%) of succeeding in the market. With the assumption that both MR and TR are reasonably independent of each other, the probability of overall project success can be expressed as

$$\text{probability of project success} = (\text{probability of market success})$$
$$\times (\text{probability of technical success})$$
$$= (1 - \text{MR})(1 - \text{TR})$$

Therefore for a project whose MR is assessed at 20%, and whose TR at 30%, the overall chance of project success is

$$\text{probability of project success} = (1 - \text{MR})(1 - \text{TR})$$
$$= (1 - 0.2)(1 - 0.3) = (0.8)(0.7)$$
$$= 0.56 \quad (56\%)$$

Even though the chances of marketing success and technical success, considered independently, are good, the changes of overall success are only slightly better than half. Certainly engineering projects undertaken deserve better odds than this.

To illustrate this idea further, suppose the MR is judged to be very high, say 90%. On the other hand the TR is judged to be very low, say 10%. What does this imbalance of risk do to overall probability of project success?

$$\text{probability of project success} = (1 - \text{MR})(1 - \text{TR})$$
$$= (1 - 0.9)(1 - 0.1) = (0.1)(0.9)$$
$$= 0.09 \quad (9\%)$$

In this case the project has less than a 10% chance of succeeding. This example shows that both risks must be acceptable—reasonably low—or the probability of project success inevitably will be low.

IDENTIFYING RISK

A biotech company that thought it had developed a cure for a certain disease was denied U.S. Food and Drug Administration (FDA) approval. Here is a case where the market risk was about zero (there was unquestionable demand).

However, even though many test trials proved positive, the FDA would not approve it because of its many undesirable side effects. And this was after $80 million had been spent.

What makes this example unusual is that the problem lay clearly with the technical risk. Most studies of new product project failures have shown that, in about 75% of the cases, the cause of the failure was due to a misjudged market.[2] This is because technical risk can be understood and managed internally by the company's own staff and technical advisers. When projects are out of control from a technical perspective, companies know what to do; they can analyze the problems, pinpoint the causes, increase the manpower, and generally manage to get the projects back on track.

Marketing risk, on the other hand, arises from outside influences that are much less subject to control. The evidence points to marketing risk as the 800-pound gorilla in projects. More projects fail because the product specification developed by marketing did not meet the customer need.[3] In marketing's defense, often marketing is supposed to provide the product specifications even though they know astonishingly less about the market or proposed product than those proposing it, who may often be from engineering. To truly understand customer needs, marketers must be deeply immersed in a market, often for years, before acquiring enough knowledge to actually serve as a customer surrogate. Unfortunately, in today's fast-moving world, many marketers do not have this level of commitment and involvement to a specific market. This is why actual real customer involvement is so important; customers provide the affirmation of the functionality needed and the financial commitment. While venturing projects can and do succeed without this assurance, customer involvement dramatically lowers the risk.

Engineering can play a critical role here before they accept the product specification. They can ask very critical questions of the persons or department that prepared the specification to see if their confidence in their assertions and plan can be shaken. They can assess what knowledge of the market the marketing team actually brings to the task: Do they have the years of experience working in the market, have they consulted acknowledged experts in the field, and

---

[2]Smith, Preston G., and Reinertsen, Donald G., *Developing Products in Half the Time: New Rules, New Tools*. Hoboken, NJ: John Wiley & Sons, 1998.

[3]Recall that this is the document created by marketing to define the product's market, features, performance, and other key characteristics for engineering.

7 MANAGING RISK IN PROJECTS

have they used as much research as seems reasonable? If the answers to these questions seem vague, it is a good time to say more information is needed.

Case Study
Disappearing Demand

Knowledgeable persons in a target company had expressed initial interest in the process of digitizing hard-copy documents to a standardized electronic format for transmission. However, the prospect company's demand dissipated during the product adaptation period. While it is often difficult to pinpoint exact reasons, the company seemed to receive price concessions on existing methods that lowered their interest in investing in the new method. Further, it appeared that internal corporate inertia to change for the better could not be overcome. While this company was just one of several that were approached and who had expressed interest, this particular target had the most initial enthusiasm and was by far the largest of the prospects.

In another situation, a large oil company expressed interest in development (by a potential supplier) of a powerful new workstation to enable the improved display of underground seismologic information. With this interest, financial commitments were made to engineer the new workstation. However, after several months, the project unraveled because key parties pushing the project at the oil company didn't have the authority to release the funds necessary to finance the project.

The Risk in New Products

Let's look at the word "new" for a moment. This is probably the most overused word in marketing. It conveys the excitement, the sense of exploration, and the desire of all persons for a fresh experience. For many the anticipation, excitement, and commitment increases when they are involved in something new. But new has a downside as well. Too often customers who expressed initial enthusiasm find a way out of their initial commitment when they realize what the often ignored risks actually are.

When customers do backtrack on their supposed commitment, the marketer often hears them using the following words. The product is

- untried
- untested
- unaccepted
- unwanted
- unadopted
- incompatible with existing methods
- only offers a small advantage
- unadaptable to our needs
- not what was desired at project outset
- not what we really wanted
- not what we saw on the drawing board
- too risky to try
- too hard to use
- too expensive
- similar to what the competition already has
- not quite right (more engineering is needed)
- not being used by anyone else
- great, but we can't use it
- great, but the finance department slashed our budgets

Excuses like these are the bane of the enterprising company and its efforts at success in the marketplace, but they are used all too frequently.

The Market Risk Matrix

Another view of the risk of developing products is depicted in Table 7.1. This provides a means of assessing risk in terms of the type of product (new or an adaptation of an existing product) and market (new or existing market for the company). This matrix assesses those projects that develop new products for new markets as highest risk, and those that develop existing product adaptations for the company's existing markets as lowest risk.

Table 7.1. Understanding Where the Market Risk Is

	existing market	new market
adapted products	low risk	moderate to high risk
new products	moderate to low risk	high risk

The principles shown by this matrix are yet another reason why it is so important for companies to protect their existing markets, the ones they know and sell most of their products to. New or adapted products made for existing markets are the lowest risk, the sweet spot of success. In direct contradiction to the phrase "No guts, no glory," most of the glory and the money can be made right in familiar territory. Venturing into new markets is truly an action fraught with peril.

Yet every day brings new companies, new products, new success: companies that overcome the odds, introduce new products to new markets and succeed. There can be a significant advantage to being first with a new product; the so-called "first mover advantage" appears to be real. Philip Kotler says, "Most studies show that market pioneers gain the most advantage. Companies like Amazon.com, Campbell, Coca-Cola, Eastman Kodak, Hallmark, Peapod.com and Xerox developed sustained market dominance."[4] How do they gain and maintain the advantage? Most often it is due to superior product performance initially, and then to the development of a strong brand name. The product becomes the standard for the new product category, as when Kleenex became synonymous with facial tissues and Xerox with office copiers.

---

[4]Kotler, Philip, *Marketing Management*, Saddle River, NJ: Prentice-Hall, 2002, p. 331.

Managing Risk

Succeeding in a risky world requires a high level of management skill. Call it what you will—smarts, luck, bravado, or skill—it all involves the astute management of risk. The way to manage market risk is to be sure that the initial specification for the desired product meets the demands of a particular set of customers (the market), and to keep those specific customers involved as the product is engineered and developed, a technique known as market assurance. There are two stages of the market assurance task. The first is to develop the product specifications to meet the customer's declared need, and the second is to continually assess the mood and commitment of the target customers as the project proceeds. Obviously, customer access and familiarity with the existing market are crucial for this approach to succeed.

Case Study
No Cure for This Project

> In a project to develop software for a supply chain application, market research showed that customers in the intended market (pharmaceutical companies) would derive strong financial benefit from the proposed software and that the software would, in many cases, improve their overall supply chain operations. The software was then developed and a marketing campaign was begun to acquire customers in the pharmaceutical market. Acquiring customers turned out to be extremely hard to do. None had bought into actually using the product during the market research phase, and none had any personal or financial stake in the software's use and success. Even though there were strong potential benefits from using the product, the software also required minor changes in business processes and cooperation between the supplier and their customers in the supply chain. The result was that use of the software and the new methods involved too much work; this project never succeeded in the marketplace. Broadly, this type of software was used in other applications in the pharmaceutical industry, but the offering company—with its sales force—was unable to gain market entry.

As remarked at the outset of this book, marketing is not easy, and one of the hardest parts is getting the continued involvement and commitment of customers throughout a project's development. If these are existing customers in

7 MANAGING RISK IN PROJECTS

existing markets, then the job is much easier because the project should be one for which they can see the obvious benefits. (Don't ever develop a project for which they *do not* see the benefits.)

Ideally the customer is pushing the vendor to develop and deliver the new capability. In this situation, the risks are technical, not marketing. As mentioned earlier, the technical side is easier to deal with. It involves engineering skill, component availability, supplier reliability, testing, schedules, and cost. These are all matters that can be challenging, but involve nowhere near the risk that a disappearing customer poses.

It's a different situation for new markets. To get up-front buy-in from new customers is a major challenge. This is why marketers who specialize in business development, especially obtaining new customers in new markets, earn the big bucks. It is a job that requires great skill in *rainmaking*.[5]

How is risk managed in these situations? A key way is to reduce or eliminate the financial risk involved. If a customer can be found who will fund all or part of the engineering development, the financial risk is minimized. This happpens most often when modifying existing products to meet special needs of existing or new customers. Then the supplier must decide if the commitment of resources to meet the special needs is worth it in terms of the firm's strategy and other opportunities or commitments.

Another way to manage risk—especially for new market and new product situations—is to manage the risk so that the adjusted risk (adjusted for expenses) actually decreases as the project proceeds and more has been spent. As a project proceeds, project management should be able to assure those financing the project (either internally or externally) that the probability of overall success is actually increasing. Management should be learning more about both the market and the engineering difficulties as time proceeds, and using this information to increase the likelihood of success. On the other hand, if the risk were unmanaged, the risk could stay the same as the project proceeds or, even worse, increase. Figure 7.1 illustrates this situation.

---

[5]Rainmaking is a term used in industry to refer to the ability of certain individuals to develop paying customers. A person skilled at this task is called a *rainmaker*. A useful book on techniques used in rainmaking has been written by Ford Harding and is entitled *Rain Making: The Professional's Guide to Attracting New Clients*, published by Adams Media Corporation, 1994.

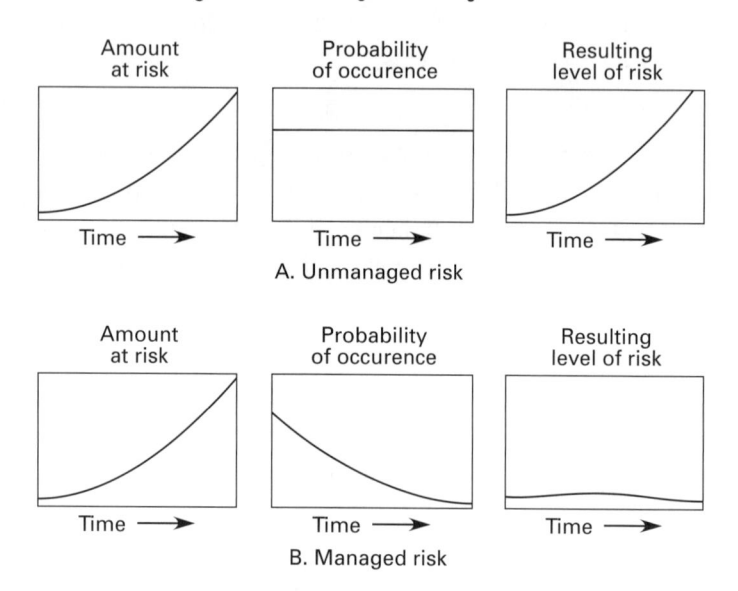

Figure 7.1. Unmanaged vs. Managed Risks

A. Unmanaged risk

B. Managed risk

Reprinted with permission of John Wiley & Sons, Inc. Smith, Preston G., and Reinersten, Donald G., *Developing Products in Half the Time: New Rules, New Tools*, copyright © 1997. John Wiley & Sons.

A real problem with this adjusted risk approach is that as more is spent, there is often more psychological and emotional commitment to the project from team members and management. Unfortunately, internal buy-in has nothing to do with whether the customer has bought in. The commitment sometimes becomes irrational. It takes a cold eye and a dispassionate heart to ask the really tough questions about future success while enthusiasm mounts internally for a project. Then it's even harder to end a project where the negative evidence appears to be mounting. The tendency of most people is to be positive and not be a naysayer.

Techniques for Exposing Risk

There should be no harbors for risk. All should be out on the table for any project. Once out in the open, risks can be managed. So a good way to begin a project is to bring out all the risks that can possibly be perceived by all the members of the project team. Each risk should be categorized (technical, marketing, supplier, competition, etc.) and assigned a place on a graph where the

7 MANAGING RISK IN PROJECTS

x-axis represents the probability of occurrence and the y-axis is the amount of money at risk. Any risk above a threshold should be subject to intense management as it represents something of high importance. Regular reviews should focus on reassessing and relocating risks on the graph and identifying new risks. Smith and Reinertsen provide an excellent review of this method in the referenced book.[6] An example is shown in Fig. 7.2.

Another method used in many companies is the so-called stage gate system. This system gets it name from the division of a project into stages, or phases. At the end of each stage there is a gate, or evaluation point, that the project

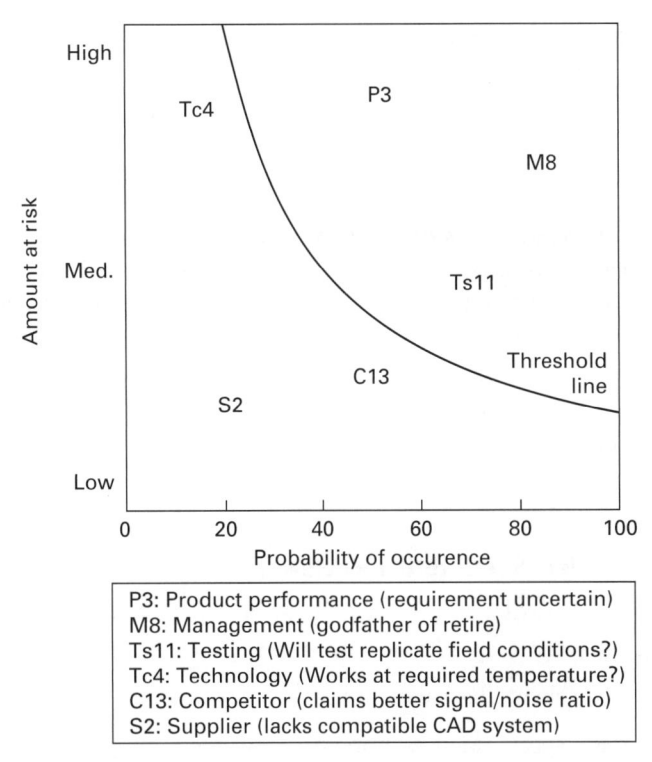

Figure 7.2. Categorizing and Assessing Project Risks

P3: Product performance (requirement uncertain)
M8: Management (godfather of retire)
Ts11: Testing (Will test replicate field conditions?)
Tc4: Technology (Works at required temperature?)
C13: Competitor (claims better signal/noise ratio)
S2: Supplier (lacks compatible CAD system)

Reprinted with permission of John Wiley & Sons, Inc. Smith, Preston G., and Reinersten, Donald G., *Developing Products in Half the Time: New Rules, New Tools*, copyright © 1997, John Wiley & Sons.

---

[6]Smith, Preston G., and Reinertsen, Donald G., *Developing Products in Half the Time: New Rules, New Tools*. Hoboken, NJ: John Wiley & Sons, 1998, p. 230.

must pass through to proceed. With this method the project is divided into 10 or so phases, starting with phases that develop the idea, marketing concept and plan. At each phase the project is reviewed by a team and a "Go" or "No Go" decision is rendered (the gate). If a No Go decision is rendered at any phase, the team members can address the deficiencies and recycle to attempt passing that phase again. Later phases focus on the engineering and testing as required for the particular project. The stage gate method provides a disciplined means of managing a project and exposing risks as the project proceeds. It has the potential to improve project success.

A criticism of the stage gate approach is that it avoids diving immediately into prototype building and review, relying too much initially on studies, research and discussion—delaying the development of a prototype that customers can actually touch, feel, and see. Many believe that developing and showing prototypes to customers very early on is the best way to obtain the customer interest and commitment required to reduce risk and increase the success rate.

Skill Builder: Assessing Risk in a Current Project

Set up a risk meeting for your project team. Brainstorm all the risks that you believe your project faces. Divide them into categories such as technical, marketing, supplier, competition, and the like. Assign a probability and a cost to each and arrange them on a graph as in Fig. 7.2.

What can your team do to confront and manage the high "Adjusted Risk" situations?

Skill Builder: Specification Review

Consider the project you are now working on.

Is there a product specification available?

If not, who should have written it?

If there is a specification, is it oriented toward customer requirements or towards technical details?

Does it allow for too much interpretation as to what the customer needs?

Does it allow for engineering flexibility in meeting the specifications?

Skill Builder: Understanding Your Market Risk

Talk with your marketing department about the risk location of the various projects undertaken by your company.

Where do they fit on the matrix shown in Table 7.1?

Are they all existing market projects?

Should there be a few new market projects just to be sure the company is challenged and extending itself to new opportunities in new markets?

Where can you find customers to back some new market projects?

8

MARCOM: The Art of Communicating with Customers

Skill Builders in This Chapter
- Positioning
- Pricing
- Promotion

Many people think of marketing as mainly creating ads for magazines, TV, and the mass media. But the crux of marketing is really about creating the company's relationship with the customer. That involves much more than a slick advertising piece. Among many other tasks, it is the job of Marketing Communications (MARCOM) to create the advertising and informational pieces that help to establish and maintain key customer relationships.

This chapter discusses MARCOM and some of the key ingredients that go into creating effective marketing materials. One of these key ingredients is positioning—how to firmly secure product placement in the customer's mind. Product, place, promotion, and price (the four Ps, known as the marketing mix) are also presented.

MARCOM

All media communications with prospective or existing customers fall under the general heading of MARCOM. Its goal is to provide information and to stimulate interest, so that the market is literally demanding the product. MARCOM stokes the channel and market with all kinds of information to build a market frenzy for a product.

MARCOM is the part of marketing that people see and hear: advertisements, product brochures, television, CD-ROMs, free pens, T-shirts and many other kinds of media and materials that communicate product features and benefits to possible customers. While a good product can go a long way toward getting the proverbial "foot in the door," the truth is that many customers need help to learn about and understand almost any product. A prospective customer base may not even know it needs your particular product; in this case, there is a need to create a market. Customers are confronted with an information barrage each day. According to Kotler, people are deluged with about 1600 commercial messages per day. Of these about 80 (5%) are consciously noted, and only 12 (less than 1%) get some reaction.[1]

Products must find an identity and potential value amidst a sea of information that continually swirls around potential customers, who have limited time to devote to the constant barrage. This deluge consists of information about

---

[1] Kotler, Philip, *Marketing Management*. Saddle River, NJ: Prentice-Hall, 2002, p. 565.

directly competitive products, as well as general information about thousands of other products, ideas, news events, sports events and other happenings.

Case Study
Understanding Market Awareness

> Here is an example of a firm that makes software that tests other software. The firm's software tests the client's software by providing inputs to the client software, assessing the results, and then recording them. The benefit of using software that tests other software is that many more variations of inputs can be applied and the results recorded than if humans were applied to the testing task.
>
> The particular market that this company addresses consists of developers of software that is used on mainframe computer systems for transaction processing. Transactions are, for example, the changes to bank accounts made by the bank's customers by adding or withdrawing their funds from their accounts. After five years in business and growing its revenue to $10 million per year, the company wanted to understand more about the market awareness of their product. They hired a market research firm to conduct a study of prospective customers. The firm reported that about half (309) of the firm's 600 prospective customers had heard of products like the one the company sold, but only six (1%) had actually bought or tried a product in the category. This was far less than what the company had originally thought, and it concluded that it needed to increase market awareness and to educate the market about its product's capabilities.

Given the world communications explosion, it may seem close to impossible to effectively get the word out about a product. And it is. Before a company can compete with another company's product's features and benefits (i.e., engage in a product war), a company needs to learn how to compete in the war of words. They must effectively describe the product to the market. This requires language that sticks in the customer's mind and is called *positioning a product*.

Positioning a Product

Fundamentally, a company needs to create and describe a place, a niche, for its product in the customer's mind. The process of finding and describing this niche is called positioning.

Effective positioning is crucial to competing in today's highly competitive markets, which are rich in products and information. Positioning seeks to distinguish a company or a product from similar companies or products. Consider a few products or companies that have had effective positioning.

- Chevrolet—the standard car for the middle-class American family (until poor quality caused it to lose that position starting in the mid-1970s)
- Walt Disney—the best, or premier, family entertainment company
- Coca-Cola—America's drink
- Intel—the highest quality chip for the personal computer
- Bell Labs—innovative originator of many communications technology advances

In each case there is almost an instant nod of agreement with respect to the above positions. People identify with them. Notice that some of the above positions were probably not consciously the work of some marketing genius expressly hired for the job. Although Intel specifically sought out the position cited above to distinguish it from clone chipmakers, the Bell Labs position was in effect conveyed on the lab through its long and illustrious history of technology innovation. It became obvious to the Intel marketing team that here was a position they should reinforce.

Once a position is established, it is very hard to dislodge. Only after many years of low-quality automobiles did the American public divest Chevrolet of its position as *the* car for the American family. Ford later won this position, beginning in the mid-1980s, with the lauded line of Taurus automobiles. Then, as the Taurus line became less popular and the American family moved on to acquire SUVs as the family car, the Ford Explorer became positioned as the SUV of choice for this audience.

The engineering behind a product plays a key role in the product's positioning, because engineering builds in the product features. If the product is intended for a technical audience, often features will be the only positioning handle available. A technical audience wants to know where a product fits among the products available, so the company must communicate to them the features that make it a unique and superior device. They must understand clearly where the product fits among other products available in the same category. The technical audience typically wants numbers as the positioning element, so-called speeds and feeds. How fast is the product, and how often or how many units will the product deliver?

The positioning message may be different for different target audiences. An executive audience, for example, looks for more general explanations of product capability, such as compatibility with the tools they already use, training needs for the new product, and the company's relationship with the product's manufacturer.

Case Study
Positioning SUN Microsystems

> When SUN Microsystems started out with its line of workstations, the operating system used was the version of UNIX known as the Berkeley distribution (BSD). For many of SUN's early customers who were technology innovators from universities and R&D labs, the BSD offered features that they were familiar with and wanted to use. SUN was recognized by this early market as a technology leader that provided an operating system they all knew and loved. SUN further catered to this market by offering a number of UNIX enhancements that were well received by the UNIX users. Based on these features and other capabilities, SUN skillfully built and solidified its position as the premier supplier of computing innovation to engineers who used workstations.
>
> This positioning effort began about 1985 and was effective until the late 1990s. SUN retained the market standing as the leading supplier of UNIX workstations to the world engineering community. Along the way its leadership was challenged by Apollo, DEC, IBM, and Hewlett-Packard, but the early work in positioning carried it through these many marketing and technical battles, well battered but retaining the crown. Unfortunately for SUN, the market's preferences for hardware swung to the Intel platform and the complementary software from Microsoft. Further, a

brand of UNIX known as Open Source UNIX became available. This combination of events eroded SUN's position as the leader in the market for technical workstations.

Fundamentally, only two important characteristics determine the position for a product. These are the product's

- differentiation
- benefits

By creating the features that differentiate the product from the competition, engineering provides the basis for positioning. Marketing emphasizes the key differentiating product factor. Customers note this one principal item and remember its benefit as the product's main advantage.

The differentiating factor with the Intel chip, for example, is Intel genuineness. No other chip can be an Intel chip. The benefit is the confidence the customer gains from not having to worry that software might be incompatible with the PC's main processing unit.

In a completely different arena, consider lawnmowers. Of the numerous brands available, consider three: Craftsman, Honda, and Toro. Table 8.1 articulates what could be the key aspects of each in formulating their positioning statement.

Table 8.1. Hypothetical Positioning Information for Three Competing Lawnmowers

brand	customer focus	key benefit	key differentiator
Craftsman	average homeowner	available at all Sears stores, backed by Sears	low price but high reliability
Honda	more dedicated lawn care homeowner	maintain your lawn more easily with better results	more intelligent mower features
Toro	for more extensive lawns	long lasting for the bigger jobs	rugged

8 MARCOM: THE ART OF COMMUNICATING WITH CUSTOMERS

With this information, the marketing professional who works for any of these companies can create a positioning statement that clearly positions the appropriate machine for the right audience. For example, for the Honda mower, the positioning statement *might* read: "For the homeowner who wants a lawn that is beautifully cut, the Honda mower brings superior technology and features that offer ease of mowing yet much better results."

Once a company agrees upon a position, the function of marketing is to turn that into a positioning statement used as the basis for the company's MARCOM. The positioning statement focuses and identifies the chief market for the product, and recites the main differentiation and benefit in easy-to-understand terms.

By using this statement as the guide for all product communications, the company both defines its market and begins to focus sharply on the main reasons a customer would buy. It also defines the position of the product in the customer's mind and makes it easy for customers to understand what the product is used for and where it could fit into their plans. Customers compare the new product with whatever they are doing at present; they try to understand how it is different from competing products and think how they could benefit from it. Then they think of buying it.

An unintended side benefit of such a positioning statement is that now the company's engineers know just who the company is targeting as customers, and what key benefits and differences they must emphasize in product development.

Product Marketing Communications

The positioning statement creates a solid foundation for proceeding in the midst of marketplace turbulence. The positioning statement guides the preparation of advertising, public relations, and product brochures so that they are consistent with one another and present a uniform message to potential customers. However, even with effective positioning, the technology adoption life cycle (see Ch. 4) must be kept in mind when preparing the types of marketing communications specific customers will read and respond to.

Table 8.2 provides an overview of the buyers in each market stage and the type of information they are looking for.

Table 8.2. Needs of Different Market Stages

market stage	buyers in this stage	what they want
early market	innovators, visionaries	new features, "speeds and feeds" information, performance information
mainstream market	early and late adopters	strong benefits, adherence to industry standards, affirmation of support, company stability
late market	laggards	low price, compatibility

By its nature, the early market does not need a wealth of brochures and advertising. These folks endlessly scan technical magazines and the trade press looking for anything new. The innovators and visionaries will readily notice technical articles, news releases, and small new product announcement ads and call for more information about the product.

Case Study
Advertising to the Early Market

Some years ago, a very small start-up company had developed a new graphics board for the personal computers of the day. Although totally unknown as a company, it ran one ad in *BYTE Magazine*, a popular consumer publication, and was deluged with calls from all over the world. The staff was not prepared to handle the inquiries or to sort out true customers with genuine needs that the product could meet from those who were just seeking literature and conversation to stay current with the industry.

So while it can be easy to get a response from early market people, the company must also have a way to quickly find customers who are willing to buy today what is available and proven. Also, there must be ways to assist those who want to work with the engineering team to adapt the new product to their needs, but not at the cost of burying the start-up operation.

In the mainstream market, the pragmatists want to buy a product with plenty of support, so they pay attention to trade shows and more traditional advertising. These customers ask questions about company longevity and reliability, and with what partners is the company affiliated. Risk aversion is a strong characteristic of buyers in this part of the market. While small ads draw in the early market, customers in the mainstream market want to see large ads that reflect the robustness of the company and product.

At this point, a company's brand identity can become an important marketing aid. Companies that have established their products under the umbrella of a strong brand name have a much easier time selling to the mainstream market. Microsoft, Intel, and HP have strong brand names. The mainstream market accepts more readily any product produced by these companies. The pragmatists and conservatives also want to see other people they know in their industry using this product. The benefits must be real and easy to see.

Customers in the late market look for the cheapest way to buy an undifferentiated, widely available product accepted in their industry. The product either fits the parameters or it doesn't. That's it.

THE FOUR Ps

To describe the key elements of product marketing, marketers often refer to the four Ps.

- product
- price
- promotion
- place

Product

As discussed previously, the *product* is described by its positioning in the market. The position is determined by a product's differentiation from other products and by its key benefit for its customers. How the product is described in marketing materials will depend on what market stage it is in.

Price

Price is an important and complicated matter. Decisions concerning the price of a product can be made on the basis of cost and other financial data, the product's presumed value to the customer, personal judgment, or any combination thereof.

Price decisions that are based on data take into account the investment required to develop the product, the variable cost of its incorporated components, and whatever company overhead may need to be supported by the product's sale. Other data to be included in the pricing decision may include competitive prices and economic trend data, such as the rate of inflation. To justify a price, sometimes a company may need to forecast the product's sales volume for a number of years (usually five), and juggle the figures to obtain an acceptable return-on-investment (ROI) figure. While in most situations it is not desirable to charge less than a cost computed from actual data concerning the product, basing pricing decisions on cost alone can exclude reasons for pricing based on the value of the product to a customer. A value-based pricing decision considers just that. It asks: "How much does the product increase the efficiency of operations, save money or time, or help in other ways that are important?" The pricing focus is on the customer need, not on the product cost. A value-based pricing decision recognizes that value is what customers get out of products, not what goes into them.

Like some decisions in marketing, the pricing decision is often ambiguous, and usually some combination of cost and value-based analysis will be used to determine the price. The price will change over the course of the product's life cycle, as it is sold into different stages of the technology adoption life cycle, so several pricing decisions must be made over the life of the product.

Case Study
Shifting Competitive Emphasis

In his book, *Competitive Strategy: Techniques for Analyzing Industries and Competitors*[1], Michael Porter makes the case that a firm can compete in only three ways: differentiation, market focus, and price. A company can

---

[1] Porter, Michael E., *Competitive Strategy: Techniques for Analyzing Industries and Competitors*. New York: The Free Press (MacMillan), 1998.

rarely use all three in its competitive strategy. However, in using the technology adoption life cycle (TALC) model of the market as a guide (see Ch. 4), the emphasis can shift from differentiation to market focus to price as the market for an innovative product shifts from early to mainstream to late. In this process, the product may indeed change so that it is properly adapted to the needs of the mainstream market, and that later, costs of production are low in order to secure a low price position in the late market.

During the early market stage, the product is new and customers will compare it with others that they buy for similar purposes. The customers will determine the relative advantage of the new product over whatever it is intended to replace. Now is the time to price the product at whatever premium the market will bear. The difference between the price and cost will, almost inevitably, be needed to fund the cost of adapting the new product to customer needs, handling the customer concerns that arise as they introduce the new product to their organization, and correcting any faults that are discovered as it is put to the test by initial customers.

As the product moves into the mainstream market, more competition will appear, causing downward pressure on pricing. Pricing can be a tactical weapon in this battle as it is used to win individual sales efforts. However, the overarching mission of the company is now to become the market leader by winning the highest market share in the product category. The market leader will amass the vast proportion of the profits to be had during the expansion into the mainstream market. Examples of companies successfully becoming dominant in the mainstream market include Oracle (databases), IBM (mainframes), Intuit (personal finance software), and Borland (C++ software development kits).

For appealing to the late adopters and laggards, pricing is a major strategic weapon because other product discriminators become less important. By this stage, the product has become somewhat generic, and less importance is attached to a particular company for offering superior value. If a company at this time has a low-cost position or is able to produce the product for less, it can have a significant edge in price.

Price is undoubtedly a key marketing issue. But it is more important, especially for new products, to establish strong product differentiation and market focus. Building unique value into the product, and focusing on markets that can most understand and use this value, can result in a defensible market position that withstands pricing attacks by even the strongest competitors.

Promotion

Promotion is synonymous with MARCOM: how companies communicate with prospective customers—through trade shows, advertising, and public relations.

For technical products, product documentation should not be overlooked as a marketing influencer. Though usually produced by the technical writing department, documentation can also be used by marketing as another form of promotion. The importance of good documentation to product marketing is easy to overlook, but when technical users evaluate a new product, the ease with which they understand it through its documentation is crucial to the product's acceptance. The product's features and benefits should be clearly described, and inexperienced buyers should readily understand the elements for using the product. The effectiveness of product documentation can increase the product's trialability and reduce its relative complexity (see Ch. 6).

Place

Finally, *place* refers to where and how the product will be sold, or in the ways the customer actually buys the product. Place is more commonly referred to as the channel (see Ch. 9), especially for engineered goods.

Skill Builder: Positioning

Consider a product you are currently working on. What is the biggest benefit of this product for its intended market?

What is it about the product that makes it superior to its competition? From this information develop a positioning statement for this product. Try out the statement on your coworkers and see if it helps them remember the purpose, function, and benefit of the product.

Skill Builder: Pricing

Consider factors that will influence the price of a product or service you are developing. Can you calculate or estimate the costs of all the contributions that will be made to this product, from the costs of raw materials and engineering to the marketing and selling costs? Given all these costs, even if only a rough estimate, how does your price compare to the prices of similar or roughly comparable products on the market today?

Is it significantly less, roughly equal, or more?

If equal or more, how can you take costs out and bring your price down?

Skill Builder: Promotion

What do you think is the best way to promote your product?

Consider a range of ways from trade shows, advertising, giveaways, internet spots, seminars, and direct mail. Discuss your ideas with the marketing department over lunch. What are they planning and why do they think it is the best way to promote?

9

Moving Products Through the Channel

Skill Builders in This Chapter
- Channel Value
- Matching Channel with Product Position in Life Cycle

One of the jobs of marketing is to choose from among many possibilities the optimum places where the product will be sold. These places are called channels. *Channel* is an appropriate word because it conveys the dynamic and flowing sense of product movement from source to customer, unlike *place*, which conveys a static sense like a neighborhood grocery store.

Selecting the channel or channels involves making the following decisions.

- where the product will be shipped to, stored, and available for sale
- how to sell the product
- whom to sell the product to
- who will support the product once it is sold

This chapter discusses these issues and relates them to the selection of channels for products that go through the technology adoption life cycle.

EXAMPLES OF CHANNEL CHOICE

Channels meet both seller and buyer needs and provide a match between the two. Depending on the type of product, the choice of channel can vary greatly, as can be seen in the three examples below.

Standardized Industrial Products

Electronic components (resistors, capacitors) and mechanical parts (nuts, washers, bolts) are often sold in high volume to selected customers. These highly standardized products are shipped directly from the factory to a major destination, such as a distributor or large original equipment manufacturer (OEM)[1] buyer. The distributor resells the product to end-user customers who incorporate it into a product. The OEM integrates the product into a product

---

[1]The term OEM, which stands for Original Equipment Manufacturer, is somewhat of a misnomer. This is because the term literally means a manufacturer of a completely assembled product composed of parts purchased from a number of sources. For example, IBM is considered to be an OEM even though its computers are composed of parts purchased from literally thousands of suppliers as well as themselves. The term VAR, or value-added reseller, usually is applied to smaller companies that integrate and assemble parts (especially software) to also provide complete products or solutions to end-user customers.

9 MOVING PRODUCTS THROUGH THE CHANNEL

it is manufacturing. In exchange for high-volume purchases, these buyers get a standard product at negotiated, favorable prices.

Factory sales representatives often work with customers to achieve a design-in into the customer's product. A *design-in* or *design-win* is the result of a successful selling effort whereby a component vendor achieves acceptance of its component as part of a product that will be manufactured in high volume. A design-in is considered a major marketing victory because it can assure a steady flow of product sales. William Davidow, in his book *Marketing High Technology*,[2] provides details of Operation Crush, a design-in campaign that helped Intel achieve a more dominant role in certain semiconductor markets. The campaign was called Operation Crush because it was intended to crush the competition, mainly Motorola. At that time Motorola had been announcing superior products to those Intel had available and was beginning to build momentum in design-in wins. Operation Crush marshaled the considerable Intel resources from engineering, sales, and marketing in a well-orchestrated campaign that emphasized Intel's current strengths as well as future products. The end result was that Operation Crush successfully positioned Intel to maintain its lead while providing time to develop new and more competitive products to match the Motorola threat.

Customizable Products

Often products must be customized en route to the final end user. Two examples of companies that must do this are Systems Integrators and Value-Added Resellers (VARs). Typically Systems Integrators assemble and support complex combinations of computer hardware, software, and networking in order to meet the requirements of an end user. VARs adapt and install software to meet the particular requirements of an end user. These channels exist because manufacturers of hardware, software, and networking products cannot meet all the myriad needs of end users. Instead, they produce products these middlemen then adapt to meet the particular and often demanding needs of end-user customers. Because they need to understand in detail the particular requirements of their end-user customers, VARs often specialize in certain markets, such as health care, retail, and manufacturing.

---

[2]Davidow, William H., *Marketing High Technology*. New York: The Free Press, 1986.

Standard Consumer Products

A third channel uses thousands of sales outlets to sell and install products bought by consumers. Mobile telephones, pagers, stereos, and many other high-tech products fit into this category. The cost of selling and delivering in this mode is high because of the lower volume of product sold to each customer. In addition, the consumer often needs product education, adding to the sales cost. The product manufacturer often gets involved directly through direct incentives to salespersons. An example of this are the "spiffs," or cash bonuses, given directly to sales reps for each unit sold. In the mobile telephone industry, cellular telephone service providers will often lower the cost of the telephone—often to nothing—because competition for subscribers is so fierce.

MAJOR ISSUES IN CHANNEL CHOICE

Choosing a channel is a crucial and strategic marketing decision. Often customer expectations dictate choosing similar channels to those of the competition. However, there is also the opportunity to develop new and innovative channels, so consider the following when plotting which channels to use.

- How much does the channel cost? A direct salesperson is expensive in terms of cost per product sold, whereas a retail outlet is optimized for low cost per transaction. What will it cost to store the product in the channel—for example, in a warehouse or at a retail outlet?

- What is the profit margin? If the margin is significant, a salesperson may be paid to explain the product as a key part of the sale. If the margin is low, then a less costly channel is needed.

- How much customization is needed to adapt the product to end-user needs? If significant adaptation is usually necessary, then a VAR or other customizing agent is required.

- What partners are needed to help complete the sale? If the product is being sold to an IBM-dominated end-user computer customer, then a partner such as a well-established consulting firm may help gain credibility and entry to key executives.

- How will the channel be informed and educated about the product?
- How will the market be stimulated to buy from the channel?
- How will the channel support the product? What are the alternatives for fixing the product when it breaks, or when the customer needs assistance?
- What is the likelihood of channel conflict? Sometimes product manufacturers authorize both VARs and their own salespersons to call on accounts. This can lead to channel conflict if both are selling to the same customer at the same time. While this practice can sometimes stimulate active and aggressive selling efforts, the customer may be confused or in some cases may even try to obtain lower prices or other benefits from the competition. Generally it is in the best interests of customers, VARs, and the product provider to minimize channel conflict.

Unfortunately, none of these choices can be made independently. Each issue influences the others, so choosing a channel in most developing markets becomes a highly dynamic situation. In the case study below, conflict arose quickly when the Apple II became a successful product. Several different channels were demanding the product, but only certain channels could fulfill what Apple thought were its customer's needs at that time in the early market. Apple therefore exercised its prerogative to sell to whom it chose.

Case Study
Early Market for Apple Computer

In the early market for personal computers, different brands were sold in a variety of ways: through Radio Shack and other stores, the mail, and a few specialized dealers.

Apple had decided to sell through retail computer dealers, usually small stores that were just beginning to be established in many cities and neighborhoods. The Apple product had a margin large enough to provide the support that the dealer would need in order to close each sale.

Given the high margin available, a number of vendors began selling the Apple via an alternate channel—mail order—advertising to customers through ads placed in the computer magazines. Mail order dealers offered significant discounts to undercut the retail dealer channel. However, they offered no support. As a result, a customer could have a local dealer provide the needed education, but avoid buying at the higher price by ordering from a nationwide mail-order discounter.

Apple quickly saw the long-term danger to their product sales and aggressively squashed this mail-order channel. The direct-mail discounters were furious since they were no longer able to fulfill orders. However, with this action, Apple's existing dealers became even more loyal and devoted to the Apple cause. Additional dealers were encouraged to carry the product, knowing that their efforts would be protected.

CHANNEL CHANGE WITH MARKET CHANGE

Just as currents and waves can change channels in bays and rivers, so do industry channels change because of the growth and change of markets. A suitable channel for an innovative product is different from that for a product that has been in the market for some time. In Ch. 4, the technology adoption life cycle (TALC) illustrates how customer characteristics and needs change as a product matures and is adopted by a market. Table 9.1 outlines how customer needs help determine channel choice in the three TALC segments.

Case Study
Channel Change

Compaq was an early manufacturer of portable personal computers. They established a dealer channel that carried Compaq products exclusively. In exchange, Compaq provided maximum support and significant margins, and established a reputation among dealers for fairness. Many individuals left careers with major firms to enter into a Compaq dealership, secure in the knowledge that there was a solid basis for receiving and supporting the Compaq product. Thus began a lucrative and productive relationship for both the dealers and for Compaq, which the company protected for quite some time.

Table 9.1. Customer Needs and Channel Functions in the Three Market Stages

market stage	customer needs	channel provides
early market	education about the product	personal selling and support
	customization of the product	flexible, adapted product with support for adaptations
	high value for customer operations	confidence in the seller and the solution
	time to decide on product and its application characteristics	staying power in the market
mainstream market	high economic gain	evidence of past success delivering economic gain
	conformation to industry standards	influence and participation in standards organizations
	references	successful previous installations
	integration with other complementary products or services	develops relationships with and recommends third-party suppliers
	success in the industry	has sold successfully to other companies in the target customer's industry
late market	low price	low-cost distribution and low margin
	mass availability	wide availability via numerous channels
	totally standard product	generic products, "plug-and-play"

However, as margins began to erode in the personal computer industry in the late 1980s and early 1990s, Compaq established an alternative channel via retail stores. Due to a downward price spiral in the industry, the dealer channel could no longer count on the margins needed to run their type of business, so they gradually disappeared from Compaq's business agenda. By this time, the PC market was changing from a mainstream to a late market, and the dealer channel started switching to networking and LAN products, products that were earlier in their life cycles. Novell, the early leader in LAN networking, was now able to provide dealers with the type of margins and support they needed.

Choosing the correct channel for the stage of the market is crucial. Attempting to sell products to a mainstream customer without delivering the product through the appropriate channel can be an exercise in frustration. Also, as a product gains acceptance in the early market, the channel must change to match the channel needs of the mainstream market.

Protecting the channel is equally important. Nothing can be worse than the bad word-of-mouth that ensues when a channel partner sees that it is being circumvented. In the case of Apple, the company terminated the mail-order channel to protect the small dealers that provided the needed education and support. At the time of the decision, the personal computer industry was at a very early stage. Amidst the considerable turmoil that existed, Apple did not know if its decision would pay off in the long-term growth of the company. The protection of this channel was not only ethically the right thing to do, but also a gamble that paid off.

Quality products, market identification and positioning, and channel choice are the elements of marketing success.

Skill Builder: Channel Value

How many channels does your product or product line use to reach the market?

Are any channels significantly more valuable than others?

Ask your marketing department if any significant channel changes are being planned.

Skill Builder: Matching Channel with Product Position in Life Cycle

Make an estimate of where several products in your company are in their technology adoption life cycle (early, middle, or late).

What channel do these products use?

Do they match up with the suggestions in Table 9.1?

10
Self Marketing

Skill Builders in This Chapter
- Your Positioning
- Your Positioning Statement

This last chapter focuses on applying marketing knowledge to career enhancement. Increased competition, more product alternatives, rapid technology obsolescence, cost advantages from country to country, and shorter product life cycles have contributed to more uncertainty and risk for companies today. The impact on engineers associated with companies subject to these pressures has, in some cases, been severe.

Layoffs, downsizing, restructuring, and other company actions affecting careers are frequently reported in the press. While there has been a large increase in defense spending because of the terrorism threat, spending for this purpose by the government puts pressure on other government programs. Thus there are fewer funds available for projects such as space and scientific programs, which employ many engineers and scientists.

Whether a new airplane, the space station, or a new electronic chip is at stake, the risk for all projects is much higher than in times past. Because engineering labor is a major component of these projects, when projects do not develop as anticipated, the risk to jobs and careers is very high.

At one extreme, career enhancement can mean maximizing growth in your present job. At the other, it can mean finding a job when necessary. It is possible to apply the key elements of marketing to finding a job when you need one.

REVISITING PRODUCTS, CHANNELS AND MARKETS

As discussed earlier, marketing involves three principal components: products, channels, and markets. For a person who wants to find the right job for his skills, talents, and education, these translate into the following.

Product

In a career search, the *product* is the person looking for a job. A convenient way of summarizing a person's features and benefits to an employer is found in the acronym KSA (Knowledge, Skills, Aptitudes). Focusing on these three attributes helps crystallize an individual's key selling points in a complex and competitive job market. It makes a strong impression clearly and quickly.

A recent engineering graduate offers a potential employer fresh skills. As workers mature, they become skilled at adapting their knowledge to real-world problems. Aptitudes and special abilities become known through testing and application on the job. The key to finding a job in a new industry lies in translating the KSAs learned in the old industry into KSAs understood and valued by the prospective employer in a new industry.

Channel

For the job market, the *channel* refers to where and how you promote yourself. This involves positioning yourself within the right industry and promoting yourself in ways that work, such as responding to job ads in the newspaper or on the internet, using recruiters, networking with acquaintances, or going to a "job-shop" for project work.

Market

The *market*, the set of possible employers, can be characterized or segmented by industry, geography, skills required, and other indicators. Often a quick review of market factors can add or eliminate companies or regions from overall consideration. For example, don't go to Seattle for a job in aerospace when Boeing is laying off. Do go to Moline, Illinois, when the worldwide demand for agricultural equipment is rising. Ignoring these kinds of market indicators doesn't mean a miracle won't happen. But why fight the odds? Go to places or explore careers where the odds are high that someone is hiring.

Besides defining product, channel, and market, the job prospect must effectively use the three ways to compete: differentiation, focus, and price (see Ch. 8). She or he must seek ways to be significantly differentiated from other prospects in the same market, focus on a market of choice, and be priced realistically in view of what the market is expecting to pay. The issue of pay often surprises engineers leaving large companies, because pay is lower in small companies, which is considered a separate market. While pride may be at stake, this is simply a market reality. Different markets possess different characteristics.

DEFINING YOUR MARKET

Like any company with a product to sell, define your market. In a competitive job market, an engineering degree must be supplemented with other KSA features and benefits.

Your market should be defined in current terms, some of which may include

- more job opportunities in small companies than large companies
- growth in project, or temporary, hiring
- fewer hardware jobs, more software jobs
- few lifetime jobs
- more government defense jobs, fewer civilian agency jobs
- fewer research jobs

Position yourself as being in the market already. This means talking to people and getting their opinions and attitudes about products, working with customers on how they are using the products, and being immersed in the dynamics of the products' availability.

This is what any person looking for a job wants to do. It is the basis for networking. Get appointments with people and knock on doors. Interview people, question them, and ask if they can suggest another person to talk to. Immerse yourself in the chosen market. The result is that you will know as much about the target market as the next person you interview with does. After a while, you may know more people and as much about what's going on in the market as the next person to see on your list. It is very impressive to be able to display an intimate knowledge of the market to a person already in it.

Markets for jobs are always expanding and contracting. Find what particular jobs are increasing in number and where they are. For example, if a person were a mechanical engineer laid off from a large aerospace company, what is that person's market? Is it only other large aerospace companies? Where are they? Are they hiring? While it is true that this person has skills other than the ones learned as an aerospace mechanical engineer, the market will position him as only that unless he can clearly explain his other KSAs in a resume and cover letter.

An engineering career is a license to a lifetime of learning. As certain skills become less needed or obsolete, fewer companies will hire on the basis of them. This doesn't mean that there won't always be a demand for COBOL programmers. But the chances of an out-of-work COBOL programmer finding an attractive position are currently much less than those of an up-to-date programmer who knows C-sharp and object-oriented programming.

DEFINING YOUR PRODUCT

As a product in the market, you will be known by your KSAs. Employers are looking for certain features and benefits that you can bring to them through your knowledge, skills, and aptitudes.

A resume is one way of expressing your KSAs to a prospective employer; a cover letter adapts the KSAs on the resume to his need. However, the job search can get very discouraging when there is a fundamental mismatch between your KSAs and the available market. This is what happens when there are large layoffs of engineers and everybody with similar KSAs is "on the street" at once. In this situation, you can either differentiate yourself from other engineers in hopes of landing one of the few jobs available, move to another location with suitable jobs, or change your KSAs to match the available market.

In any case, you confront an increasingly competitive market with more job seekers chasing fewer jobs. In such an environment, job seekers should pay attention to methods for competing with other job seekers. As seen earlier, there are essentially only three ways to compete: differentiation, focus, and price. Here are some suggestions.

- *Differentiation:* How is one engineering candidate different from the others in terms of accomplishments, projects, education, or outside activities? Pick one point of relative advantage and emphasize it in order to stand out in the crowd.
- *Focus:* Define your market and understand what its needs are. If looking for a position in the same or similar industry, use market knowledge to advantage. If that market is saturated or not hiring, research a new market. The best way to do research is by networking. Use the

phone, talk to personal contacts, attend trade shows, or join a professional association to expand your knowledge and references in the target market.

- *Price:* This is mainly salary, but may also include such benefits as relocation allowances, sign-up bonuses, benefits, and the like. Salaries are cyclical; they are more or less dependent on the local, industry-defined market. So the best policy is one of flexibility.

CHOOSING THE CHANNEL

After defining your product and choosing the target market, you need to choose a channel to link yourself to the market. Networking, as mentioned above, is a powerful tool for conducting research about the market, but it also serves as a channel because the contacts made in this activity can leverage you into the market.

Positioning is an important channel activity. As mentioned earlier, the key elements of positioning are the benefit you can bring to the market, and the differentiation that makes you stand out from the other candidates. The candidate should use whatever makes him unique and stand out from the competitive crowd.

Drake, Beam, and Morin (DBM), one of the leading outplacement specialists, recommends rehearsing a two-minute "elevator pitch" that summarizes background, career, and accomplishments. While this short speech is valuable, it should conclude with a personal positioning statement. So, after giving a brief overview of your education and career history, you should immediately pitch the benefits you can bring to the interviewer's organization, and what is distinctive about you. Said with conviction, the listener should immediately understand your value to him and why you are different from the other candidates.

Paraphrasing the earlier positioning statement (see Ch. 8), here is how it might sound.

"...Therefore, for companies that make your kind of widgets, Mr. Jones, (*focus*) I bring 12 years of experience in ensuring that these widgets are manufactured to international specifications, which would be very important for a company looking to expand abroad (*benefit*). Besides this benefit to your company, I have several years of international experience in assuring that the widgets are supported in numerous foreign countries (*differentiation*)."

If you have chosen your positioning statement to meet the needs of the market, it should leave a positive, lasting impression in the interviewer's mind.

Finally, price must be right for the market. In some cases, the hiring company will want to know all previous salary details up front. Avoid revealing this information, if possible, because most companies will gauge salary negotiation on that information. If you have created a distinctive position with the prospective employer, you should, in fact, be in the position to demand a premium salary because you have created the idea of demand through differentiation and focus.

Skill Builder: Your Positioning

Consider your own positioning in the market for jobs in your general skill area.

How do you differentiate yourself from other candidates?

What benefits do you bring to the company hiring you?

How will you focus your marketing efforts?

Skill Builder: Your Positioning Statement

Put together your own positioning statement. Try it out on close friends and see what they think. Ask if they think that it reflects your own unique talents and the special something or benefit that you bring to an employer.

References

Buzzell, Robert D., and Bradley T. Gale, *The PIMS (Profit Impact of Market Strategy) Principles*. New York: The Free Press (Macmillan), 1987.

Christensen, Clayton M., *The Innovator's Dilemma*. New York: HarperBusiness Essential, 2003.

Christensen, Clayton M., and Michael E. Raynor, *The Innovator's Solution*. Cambridge, MA: Harvard Business School Press, 2003.

Davidow, William H., *Marketing High Technology*. New York: The Free Press (Macmillan), 1986.

Drucker, Peter, *Management*. New York: Harper & Row, 1974.

Fabris, Richard H., *A Study of Product Innovation in the Automobile Industry During the Period of 1919–1962*, University of Illinois (Urbana), PhD Thesis, 1966.

Foster, Richard N., *Innovation: The Attacker's Advantage*. New York: Summit Books, 1986.

Grady, Robert B., *Practical Software Metrics for Project Management and Process Improvement*. Englewood Cliffs, NJ: Prentice-Hall, 1992.

Hanan, Mack, *Successful Market Penetration: How to Shorten the Sales Cycle by Making the First Sale the First Time*. New York: AMACOM, 1987.

Harding, Ford, *Rain Making: The Professional's Guide to Attracting New Clients*. Cincinnati, OH: Adams Media Corporation, 1994.

Kotler, Philip, *Marketing Management*. Saddle River, NJ: Prentice-Hall, 2002.

Lennon, John, and Paul McCartney. "When I'm Sixty-Four," sung and recorded by the Beatles, 1969.

McKenna, Regis, *The Regis Touch: Million-Dollar Advice from America's Top Marketing Consultant*. New York: Addison-Wesley, 1985.

Moore, Geoffrey A., *Crossing the Chasm*. New York: HarperCollins, 2002.

Perucci, Robert, and Joel Gerstl, *Profession Without Community: Engineers in American Society*. New York: Random House, 1969.

Pisano, Gary P., and Steven C. Wheelwright, "High-Tech R&D," *Harvard Business Review*, September-October, 1995.

Porter, Michael E., *Competitive Strategy: Techniques for Analyzing Industries and Competitors*. New York: The Free Press (Macmillan), 1998.

Rogers, Everett M., *Diffusion of Innovations*. New York: The Free Press (Macmillan), 2003.

Schlitz, Don, "The Gambler," sung and recorded by Kenny Rogers, Hal Leonard publisher, 1978.

Smith, Preston G., and Donald G. Reinertsen, *Developing Products in Half the Time: New Rules, New Tools*. Hoboken, NJ: John Wiley & Sons, 1998.

Utterback, James M., *Mastering the Dynamics of Innovation*. Cambridge, MA: Harvard Business School Press, 1994.

Index

A

Adjusted risk, 89
Adopters
  -early, 40 (fig)
  -late, 40 (fig)
Advertising, 101–103
Analyzing, 35, 36
ANSI, 53
Apple, 16, 17, 64, 66
  iPOD, 16
  -mini, 44
  -nano, 44
Assembly line, 13
Attacker, 54
Auto industry, 50, 51, 52 (fig)

B

Backhoe, 59, 60
Beachhead, 30
Benefits, 100
Bowling analogy, 30, 31
Brand identity, 100
Business development, 89

C

CAD/CAM, 23
Channel, 10, 11, 106–116
Chasm, 36, 43, 44

Christensen, Clayton, 27, 28, 57–60
Compaq Computer, 18, 115
Compatibility (diffusion factor), 78 (tbl)
Complete solution, 43
Customer relationships, 10
Customer, creating, 5

D

Delphi technique, 38
Deming, W. Edwards, 67
Design-in, 111
Design-win, 111
Development time, 64, 70
Differentiation, 29, 100, 105
Diffusion of Innovations, 78
Digital Equipment Corporation (DEC), 7, 8, 12, 18, 56
Disk drives, 58, 59
Disruption, 60
Disruptive, 60
Distribution, 110
Dodge auto, 51, 52
Dominant design, 50
Drucker, Peter, 4

E

Early adopters, 40 (fig), 41
Eastman, George, 34

EDI, 42
Edison, Thomas, 34
Electronic Data Interchanges, 42
Engineering,
  communication, 10
  concurrent, 14, 15
  model, 11 (fig)
  process, 13

F
FDA, 83, 84
Firestone, 3, 4
Focus, market 29
Focus groups, 37
Ford Motor Company, 3
Forecasting, 35, 36
Foster, Richard N, 53
Four Ps, 96, 103

G
Generics, 27
GKS, 76
Graphical kernel system, 76
Green space, 30

H
Harman International, 18
HIPAA, 52, 53
House of Quality, 68, 69 (fig), 75

I
Innovations, diffusion of, 78 (tbl)
Innovators, 40
Intel, 111

J
Job to do, 27, 28, 29 (case study)
Jobs, Steve, 17
Juran, J.M., 67

K
KSA, 118
Kotler, Philip, 87

L
Laggards, 40 (fig), 44
Later adopters, 40 (fig), 43
Lawnmowers, 100
Life extension, 43

M
Macintosh, 44
MARCOM, 96, 101
Market
  assurance, 88
  awareness, 97
  early, 39, 40 (fig), 50
  growth, 50
  late, 39, 40 (fig)
  mainstream, 39, 40 (fig), 46
  matrix, 86, 87 (fig)
  need, 26
  pulse, 15, 16, 17
  pioneer, 87
  risk, 82–84
  segment, 22
  stages, 102 (tbl)
Marketing, defined, 1, 22
  model, 11 (fig)
Marketing mix, 96
Marketing Science Institute, 66
Matched filter, 10
McKenna, Regis, 19
Minicomputers, 58
Money pot, 58
Moore, Geoffrey, 38, 43

N
Networking, 120
New, 85
Noise, 10, 11
Non-consumption, 29

O
Observability, 78 (tbl)
OEM, 110
Olsen, Ken, 18
Operation Crush, 111

P
Pharmaceutical, 27
PIMS study, 66
Place, 96, 106
Poker, 82
Porter, Michael, 29
Position, positioning, 97-99
Price, 96, 104
Product, 96, 103
  life cycle, 72, 73 (fig)

management, 7
  risk, 91 (fig)
Productivity, 64, 70
Product-life extension, 77
Promotion, 96, 106

Q
QFD, 68
Quality, 64–69
Quality Function Deployment, 68

R
Relative
  advantage, 46, 78
  complexity, 78 (tbl)
Resume, 121
RISC, 55, 56
Risk, 3, 83, 88–92, 90 (fig)
Rogers, Everett, 39, 78

S
Scenario analysis, 38
S-curve, 53, 54, 60
Segment, 22, 23
Segmentation, 24, 25 (tbl), 26 (tbl)
Self-marketing, 117-123
Six Sigma, 65
Software, bugs, 16
Sony, 28, 35
SPARC, 55, 56
Stage-gate system, 91
Strikes, analogy, 30
Sun Microsystems, 55, 99
Sustaining, 60
System integration, 111

T
TALC, 39, 40 (fig)
Technical risk, 82, 83
Technology Adoption Life Cycle, 39, 40 (fig), 101, 104
Total Quality Management, 68
TQM, 68
Transistor, 28
Trend analysis, 36
Trialability, 78 (tbl)

U
Utterback, James, 51

V
Value Added Reseller (VAR), 111
Value, 72, 74
  chain, 60
  network, 58, 59 (fig)
VAX, 55, 56
Visionaries, 40 (fig), 41
Voice of the customer, 65, 68, 69 (tbl)

W
Weed Eater, 29
Whole product, 41
Workstations, 12

Z
Zero defects, 65

Turn to PPI for Professional Growth
Increase Your Professional Value
Visit www.ppi2pass.com today!

Beyond Engineering:
How to Work on a Team
Suzanne Young and Harry T. Roman

An introduction to working as a productive member of a team, for engineers and other technical professionals.

Learn how to
- Bring together different personalities
- Communicate effectively
- Manage conflict

Marketing Fundamentals for Engineers
Stan Haavik

An introduction to how products are marketed and sold, for engineers and other technical professionals.

Learn how to
- Develop marketable products
- Identify customer needs
- Create new markets
- Work effectively with marketing professionals

Engineering Your Start-Up:
A Guide for the High-Tech Entrepreneur
James A. Swanson and Michael L. Baird

A complete guide to launching and growing a successful high-tech company, for engineers and other technical professionals.

Learn how to
- Manage the 5 critical elements of a successful start-up
- Write a winning business plan
- Secure start-up funding
- Protect your intellectual property

Getting It Across:
A Guide to Good Presentations
Carole M. Mablekos

An introduction to making presentations in a business environment, for engineers and other technical professionals.

Learn how to
- Plan your presentation
- Select and prepare your visual aids
- Improve your delivery
- Avoid common pitfalls

Work with Anyone Anywhere:
A Guide to Global Business
Michael B. Goodman

An introduction to international business communication and other business issues across national and cultural boundaries.

Learn how to
- Understand international business principles
- Recognize and adapt to important cultural differences
- Negotiate, market, and manage effectively in a global environment
- Work successfully within specific regions and cultures

Creative Product Development:
From Concept to Completion
Michael J. Dick

A practical guide for technical professionals participating in the product-development process.

Learn how to
- Spark creativity in yourself and your product-development team
- Reduce product-development risk and effort by starting with great ideas
- Capitalize on new product opportunities before they are lost to competitors